WHAT YOUR COLLEAGUES ARE SAYING . . .

The most effective secondary mathematics teachers are likely to have a ragged, well-loved copy of *Answers to Your Biggest Questions About Teaching Secondary Math* on their desks. This book synthesizes decades of research that supports strong student learning outcomes and distills it into practical, easy-to-implement recommendations for teachers. New and experienced teachers alike will find new and important ideas that can be used immediately to strengthen their practice and increase engagement with their mathematics learners.

Michael D. Steele
Past-President, Association of Mathematics Teacher Educators
Executive Council, Conference Board of Mathematical Sciences
Zionsville, IN

Whether you are an aspiring, beginning, or experienced teacher of mathematics, the authors have put together an indispensable book that will become your most dog-eared professional resource. Combining findings from the literature with their extensive experience as teachers, the authors answer some of the most frequently asked questions in five critical domains of effective mathematics teaching. This is the resource I wish I had when I was teaching!

Matt Larson
Past-President, NCTM
Lincoln, NE

This book answers key questions essential to setting up a classroom environment where students thrive using practical examples, tips, suggestions, and addressing frequently asked questions. This is a great handbook for any new teacher trying to figure it all out.

Chonda Long
Director of Professional Development, NCTM
Springfield, VA

Answers to Your Biggest Questions About Teaching Secondary Math provides an excellent resource for new math teachers and veteran teachers alike. The advice is practical and written such that strategies for increasing access and equity in mathematics can be implemented at any time of the year. The book centers students as doers of mathematics and gives the reader easy-to-read advice for how to promote a discourse-centered classroom where every student can be successful in mathematics.

Emma Vierheller
High School Mathematics Teacher, Penta Career Center
Perrysburg, OH

The authors of *Answers to Your Biggest Questions About Teaching Secondary Math* are to be commended for writing this important book. How I wish I would have had such a resource when I began my career as a high school mathematics teacher. This book covers the why and the how of creating equitable mathematics classrooms from the very beginning. I also appreciate the strong focus on agency and identity. This book is a must-have for new secondary mathematics teachers. Experienced teachers will also find a wealth of resources for improving mathematics instruction.

Kyndall Brown
Executive Director
California Mathematics Project Statewide Office
University of California, Los Angeles, CA

I highly recommend *Answers to Your Biggest Questions About Teaching Secondary Math* for current teachers, preservice teachers, instructional leaders, and administrators as a guide to embodying foundational truths about mathematics, including that every student is capable of developing deep mathematical understanding and contributing to the knowledge of the classroom community. Readers will be drawn to the insightful tips, concrete content examples, and helpful tables, which illuminate strategies and beliefs to thrive. Add this book to your shelf today!

Sarah Bush
Professor, K–12 STEM Education
Program Coordinator, PhD in Mathematics Education
Board of Directors, NCTM
Winter Park, FL

Answers to Your Biggest Questions About Teaching Secondary Math is a must for every teacher of mathematics! This is a resource filled with practical guidance, accessible resources, and opportunities to reflect on your teaching journey. Whether you are just starting your journey or far into that journey, keep this book with you throughout and enjoy the journey!

Trena L. Wilkerson
President, NCTM 2020–2022
Waco, TX

Answers to Your Biggest Questions About Teaching Secondary Math is a fabulous tour de force of the big ideas in teaching mathematics that will enable preservice and in-service teachers to optimize their teaching practices and to maximize the opportunity to learn for all students. It is the most practical condensation of recommendations for the teaching of mathematics to date.

J. Michael Shaughnessy
Past-President, NCTM 2010–2012
Portland, OR

Answers to Your Biggest Questions About Teaching Secondary Math is an invaluable distillation of the most current, research-based best practices with practical strategies for implementing them. Written by classroom teachers, for classroom teachers, this book belongs on the shelf of every secondary math teacher.

Jason Slowbe
High School Math and Computer Science Teacher
Great Oak High School
Temecula, CA

Answers to Your Biggest Questions About Teaching Secondary Math is a must-have for every new and early career teacher and a great resource for those who are experienced. The wealth of knowledge and strategies shared here make this an invaluable resource and will positively impact student learning. I wish I had this when I started teaching middle school 25+ years ago!

Kevin Dykema
Eighth Grade Math Teacher
Mattawan Middle School
President-Elect (2021–2022), NCTM
Mattawan, MI

This book is incredibly helpful to understand key ideas and practices for new international teachers. The book gives a very efficient and easy-to-understand overview of critical questions a teacher can have while teaching high school mathematics. As a mathematics educator, I would definitely recommend this book to my senior mathematics teachers, reading and reflecting on the questions in the book will support their readiness and transition into being a teacher.

Zuhal Yilmaz
Graduate School of Education
University of California, Riverside, CA

Answers to Your Biggest Questions About Teaching Secondary Math employs a strengths-based approach to promote high-quality secondary mathematics teaching. The authors draw from a wealth of experiences to offer practitioners activities and resources to engage students in the mathematics teaching and learning dynamic. This book is an invaluable resource for advancing accessible approaches to bolster student agency in mathematics classrooms.

Christopher C. Jett
Professor, Mathematics Education
University of West Georgia, GA

Answers to Your Biggest Questions About Teaching Secondary Math provides guidance to middle and high school teachers who are trying to change the nature of the way math is taught and learned in their classrooms. From how to build a classroom community to how to assess and advance student learning, the five key questions focus on essential elements required to create an environment in which each and every student can flourish. The practical advice provided reflects the cumulative experience and wisdom of the author team and is sure to resonate with teachers.

Margaret (Peg) S. Smith
Professor Emerita and author of *Five Practices for Orchestrating Productive Mathematics Discussions*
Pittsburg, PA

Teaching is complex! This book helps break that complexity into manageable approaches. As someone who has been leading professional development for over 20 years, this book helped me consider new and important ways to think about common questions and dilemmas teachers encounter every day. If your focus is on developing a more equitable classroom then this book is for you!

Jen Mossgrove
Director, Teaching Fellows Program
Knowles Teacher Initiative
Moorestown, NJ

This book does an excellent job formalizing a comprehensive approach to creating an environment in the mathematics classrooms that promotes equity, access, and a voice to all learners. I highly recommend this book to all mathematics teachers—beginning and experienced—and teacher educators, as it provides a blueprint for effective mathematics instruction.

Nadia Monrose Mills
Assistant Professor of Mathematics
Director of the STEM Institute
University of the Virgin Islands
St. Thomas, VI

ANSWERS *to Your*

BIGGEST QUESTIONS *About*

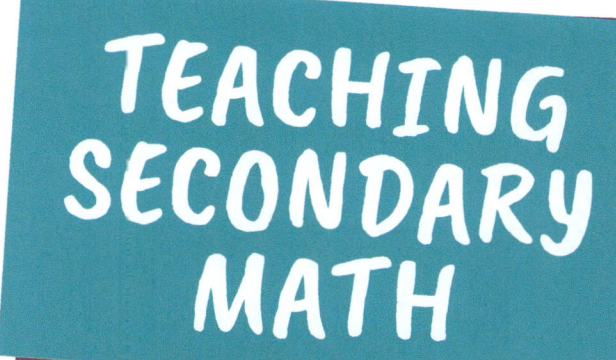

TEACHING SECONDARY MATH

FIVE to THRIVE

ANSWERS *to Your*

BIGGEST QUESTIONS *About*

TEACHING SECONDARY MATH

Frederick L. Dillon

Ayanna D. Perry

Andrea Cheng

Jennifer Outzs

CORWIN

For information:

Corwin
A SAGE Company
2455 Teller Road
Thousand Oaks, California 91320
(800) 233–9936
www.corwin.com

SAGE Publications Ltd.
1 Oliver's Yard
55 City Road
London, EC1Y 1SP
United Kingdom

SAGE Publications India Pvt. Ltd.
B 1/I 1 Mohan Cooperative
Industrial Area
Mathura Road, New Delhi 110 044
India

SAGE Publications
Asia-Pacific Pte. Ltd.
18 Cross Street #10–10/11/12
China Square Central
Singapore 048423

President: Mike Soules
Vice President and Editorial Director:
   Monica Eckman
Publisher: Erin Null
Content Development Editor:
   Jessica Vidal
Editorial Assistant: Nyle De Leon
Production Editor: Tori Mirsadjadi
Copy Editor: QuADS
   Prepress Pvt. Ltd.
Typesetter: Integra
Proofreader: Susan Schon
Indexer: Integra
Cover Designer: Gail Buschman
Marketing Manager:
   Margaret O'Connor

Library of Congress Cataloging-in-Publication Data

Names: Dillon, Frederick L., author.
Title: Answers to your biggest questions about teaching secondary math / Frederick L. Dillon, Ayanna D. Perry, Andrea Cheng, Jennifer Outzs.
Description: Thousand Oaks : Corwin, 2022. | Series: Five to thrive | Includes bibliographical references.
Identifiers: LCCN 2021057203 | ISBN 9781071870792 (paperback) | ISBN 9781071870808 (epub) | ISBN 9781071870815 (epub) | ISBN 9781071870839 (adobe pdf)
Subjects: LCSH: Mathematics--Study and teaching (Secondary) | Mathematics teachers--Training of.
Classification: LCC QA11.2 .D55 2022 | DDC 510.71/2--dc23/eng/20220120
LC record available at https://lccn.loc.gov/2021057203

This book is printed on acid-free paper.

22 23 24 25 26 10 9 8 7 6 5 4 3 2 1

CONTENTS

ACKNOWLEDGMENTS

As a team, we would like to acknowledge and thank the many math teachers, leaders, researchers, and partners who have helped shape our understanding of what teaching high-quality mathematics can and should be. We are thankful to the students we have had the honor of working with, the teachers who have worked with us as colleagues, and the professional learning communities we have shared with, and for the teachers we have coached and mentored as we joined together for our mutual growth. We thank all those who give back to our professional organizations, creating opportunities for continued learning and connecting. We thank the Corwin team for their hard work, partnership, problem solving, creativity, and vision for how to help teachers. A special thank you is necessary for the author team for the K–5 *Five to Thrive* book as they set the stage for what we are able to do. The unique challenges of teaching with the realities of the COVID-19 pandemic strengthened our beliefs that educators go the extra mile for their students every day, under any conditions. We are inspired by them. We look forward to a new normal that rekindles the best practices of the past, builds on the successes of distance and hybrid learning (and there were many!), and looks to do even more for each and every student.

Fred would like to give special thanks to his mentor, Margaret (Margie) Raub Hunt, for encouraging him with support and opportunities. He would also like to thank Margaret (Peg) Smith and Chonda Long, good friends who have helped him grow as a mathematics educator. Additionally, Fred is thankful for his many students and colleagues from his many years of working in mathematics education. Every day with them has provided new learning and understanding. Fred is especially thankful for this author team. The team built a community that shared ideas, hard work, support, and friendship. Fred also has a special acknowledgment for the team's editor and publisher, Erin Null, for the opportunity to work on this book and for her support, questions, and suggestions.

Ayanna would like to thank God from whom all blessings flow. She would like to thank Fred, Jen, Andrea, and Erin for their generous exchange of ideas and truly collaborative working style. She thanks her husband for caring for their three daughters while she worked on this manuscript. She is thankful to all of her Knowles colleagues and math friends, with special thank yous to Zuhal Yilmaz and Jennifer Mossgrove, two math educators who have always pushed her thinking and helped her cement her ideas. A special thank you to her parents, sisters, and sister-friend Nadia Monrose who prayed, believed, and rejoiced at every opportunity. Finally, thank you to the teachers from whom she continues to learn every day.

Andrea would like to thank her husband and family for their continuous support and guidance. She is also grateful to her incredible colleagues and students who continue to help her grow and learn as an educator. A special thanks to the Knowles Teacher Initiative that has been fundamental to her growth, especially in the first few years of teaching. Writing this book with Fred, Ayanna, and Jen has been an honor and joy. Thank you all!

Jennifer would like to thank all the amazing students she has had the honor of teaching and colleagues she has worked beside as well as her family for their support. She is grateful for the PCMI (Park City Mathematics Institute) Teacher Leadership Program and NCTM for not only providing exceptional professional development, but also for the lifelong friends that have pushed her thinking. A special thanks to the Fab Four, her tribe of crazy math friends, and Gail Burrill, who she is always learning from. She would also like to thank Fred, Andrea, and Ayanna for being an amazing group of educators and a pleasure to work with on this writing adventure.

ABOUT THE AUTHORS

Frederick L. Dillon is an author and national mathematics curriculum and professional learning consultant based near Cleveland, Ohio. He is a retired teacher with 35 years of classroom experience, having taught grades 7 through college. Fred is a frequent speaker at national, state, and local conferences and institutes. He is active at the local, state, and national levels in professional organizations, including serving on the board of directors for the National Council of Teachers of Mathematics (NCTM). Fred was a recipient of the Presidential Award for Excellence in Science and Mathematics Teaching and the Christopherson-Fawcett Award from Ohio Council of Teachers of Mathematics for Lifetime Contributions to Mathematics Education.

Ayanna D. Perry, PhD, is an associate director for the Teaching Fellows Program at the Knowles Teacher Initiative. There, she and her team of experienced mathematics and science educators mentor, coach, and facilitate professional development for secondary mathematics and science teachers across the nation. Prior to this role, she was a classroom teacher and an adjunct instructor for the Penn Residency Masters in Teaching Program at the University of Pennsylvania. She is an author and presenter.

Andrea Cheng is a 9–12 math teacher in Union City, California. She has also taught in Oakland at a high school for recently arrived immigrants. Andrea's passion for teaching is deeply rooted in her family experience and community. Her parents emigrated from Mexico and "served as strong, hard-working role models making [her] appreciate, value, and pursue education to [her] fullest capacity." Andrea earned a bachelor's in mathematics, teaching credential, and master's in math education from the University of California, Berkeley. She has presented at Teachers for Social Justice Conference on "Preparing students for college-ready mathematics through dual-enrollment."

Jennifer Outzs has taught for 32 years as a middle school mathematics teacher in both private and public schools in Ohio and Florida. Jennifer is currently serving on the board of directors for NCTM and was previously on staff at the PCMI (Park City Mathematics Institute) Teacher Leadership Program. She has worked as a facilitator for NCTM's Effective Teaching with Principles to Actions Institutes, served on program committees for NCTM annual and regional conferences, worked as a mathematics review panelist for College Board, and has presented at loca , state, and national conferences.

There's a common saying among veteran teachers that,

> ## I wish I could go back and teach my first year knowing what I do now.

The quote reflects the reality that the first few years of teaching are a learning experience not only for your students but also for the teacher. This is even true for those who have been teaching for a while because every time you change schools, grades, or topics, or sometimes just as your students vary from year to year, you learn and grow. Every year you teach, your experiences help your craft evolve. As you progress, you learn about your curriculum and the trajectories embedded in it, about how to make your classroom a safe and welcoming space, and about how to differentiate instruction. You will be amazed as your understanding of topics you thought you knew well deepens because you are looking at them through your students' eyes. This book shares the things we wish we had known when we started and the important things we've learned as we have improved our practice.

WHY IS TEACHING MATH DIFFERENT TODAY?

Traditional mathematics teaching may have served learners living in the Agricultural, Industrial, and Information Ages, because employment opportunities rewarded people who could think in systematic ways and refer to set approaches to solving problems. That changed as we entered the Conceptual Age and postsecondary employment sought out people who could think critically, solve problems, and demonstrate the transfer of ideas in new settings (Wathall, 2016). We've now progressed past the Conceptual Age, and employers are looking for employees who can think critically and creatively, solve problems, describe problem spaces and transfer ideas to new settings and recognize patterns and structures that define the new settings to support more efficient transfer and sensemaking. Teaching conceptually can support students in engaging creatively outside of school because concepts, by definition, are "mental constructs, which are timeless, universal, and transferable across time or situations" (Wathall, 2016, p. 6). Answers to mathematics problems that can be found by entering problems into computational technology do not contribute to creative student thinking. While learning mathematics facts and procedures has its place, students need to be taught math in ways that support them in understanding how concepts, facts, and procedures are connected so they can retain this knowledge and build a more complex schema of mathematics.

Future employment opportunities are not the only reasons to teach mathematics differently. Decades of research have added to educators' better understanding of how the brain learns and makes sense of experiences and how teaching can contribute to more connected and lasting learning. As educators, we are still learning and contributing to this body of knowledge in ways that might change some of what we currently know and understand years into the future. That is, some of the questions we answer in this book will likely be answered differently in 20 years. Though we expect to learn more about the work of teaching, much of what we know now will likely *not* change because the positive impacts of student-centered instruction and a focus on thinking and reasoning are mainstays in secondary mathematics education.

As teachers, we need to help students see mathematics as a way to make sense of the world, make predictions about future events, solve interesting problems, and draw conclusions based on data. We teach so that students may experience mathematics and mathematics classrooms as spaces where innovation, creativity, communication, wonder, and beauty regularly take place. For this to be the common experience of our students, we must teach math very differently than the ways we have learned it in the past. The eight mathematical teaching practices (National Council of Teachers of Mathematics [NCTM], 2014, p. 10) describe components of effective mathematics teaching that lead to deep and lasting understanding of mathematics for students. These practices are discussed in depth in the book, but the table below shares the practices and a brief description of each. The practices that are most closely related are color coded. While there are practices that are more tightly connected, all of the practices are useful to support effective teaching.

Practice	Effective math teaching . . .
Establish mathematics goals to focus learning	begins with clearly defined learning goals that connect concepts, procedure, and skills. These goals are connected to course standards and situated within curricular and unit plans.
Support productive struggle in learning mathematics	provides students with the support they need to persevere, communicate their progress and challenges and their emergent understandings while working on worthwhile mathematical tasks.
Implement tasks that promote reasoning and problem solving	requires the use of worthwhile tasks that provide students opportunities to engage in sensemaking. These tasks should be accessible and have multiple solution pathways.
Use and connect mathematical representations	supports students in engaging in sensemaking across multiple representations and solutions paths developed from engaging in worthwhile tasks.
Build procedural fluency from conceptual understanding	supports students in building procedural fluency by engaging in worthwhile tasks that foreground conceptual understanding. Over time, this supports flexible problem solving for students across various forms of tasks.
Facilitate meaningful mathematical discourse	supports students in engaging in mathematics discourse by sharing strategies and structure to teach students how to share their ideas and facilitates that discourse to support learning. A major component of the discussion is the multiple representations that students share.
Pose purposeful questions	uses questioning to highlight, extend, challenge, and connect student reasoning.
Elicit and use evidence of student thinking	gathers and uses evidence of students' thinking to assess students' understanding and make adjustments to instruction. This evidence could be verbal but can also include written artifacts.

WHAT EXACTLY *IS* DIFFERENT ABOUT TEACHING AND LEARNING MATH TODAY?

When students enter middle and high school mathematics courses, the abstraction of math concepts greatly increases. This abstraction has been the cause of exasperated looks and heavy sighs on the part of caregivers when their students ask for help on homework or assignments. Traditional ways of teaching that focus on memorization have resulted in caregivers who have tenuous memories of what they learned in their math courses. The ways of teaching championed by this book aim to change how students learn and the likelihood that students will develop deep understanding that remains with them into adulthood.

Today, mathematics instruction

- focuses on conceptual understanding, procedural fluency, critical thinking, and transferability of mathematics understanding to real-life problems;
- encourages collaborative problem solving;

- invites students' communication of thinking, justification, reasoning, and questioning;
- and incorporates multiple representations, solution pathways, and answers into lessons to support student sensemaking.

Though this approach is not universal across U.S. classrooms, research and practice support the use of more conceptually based instruction and recognize its benefits. The table below shows the difference between how mathematics has been positioned in the past and how mathematics instruction positions mathematics now.

Math then	Math now
Focused on speed and recall.	Focused on efficiency, flexibility, and sensemaking.
Working alone.	Working collaboratively to reach mathematical understanding.
Applying a singular procedure/strategy modeled by the teacher or text.	Using common procedures with understanding and/or developing a unique solution path to worthwhile tasks. Choosing flexibly from different approaches to solve a problem.
Focused on finding and sharing the correct calculations.	Focused on the connections between calculations, concepts, and procedures. Invites reasoning about connections.
Focused on learning how to employ basic skills.	Focused on developing conceptual understanding.
There are "math people."	All people are capable of learning math and engaging in sensemaking.
Math problems that lacked context "Real-world" problems that require students to suspend their lived experience to find expected solutions. Unrealistic "story problems"—e.g., age problems or train distance/rate problems.	Allows the application of outside-of-school realities in seeking problem solutions. Tasks are based in relevant contexts.

When comparing math taught in traditional ways with math today, you can see that math is taught to focus on conceptual understanding and not solely procedural fluency and memorization. While the need for students to factor quadratics, simplify radicals, and apply the unit circle to trigonometry problems is still important, developing an understanding of the concepts that undergird these skills is not only more important but aids in recalling and applying them when needed. In addition, through engaging in these conceptual tasks, students develop as *doers* of mathematics. Doers solve problems, develop models, communicate their reasoning, and engage in sensemaking.

How our society thinks about what math is has been shifting for some time. What it means to do math, who is expected to be successful at math, and why mathematics matters have been changing over time, though these shifts are difficult for some to see. Effective mathematics teaching upends faulty beliefs about mathematics and reinforces the truth about mathematics.

The truth about math	Faulty beliefs
Everyone uses mathematics and can use mathematics to make sense of the world around them and illuminate inequities that affect them.	Math is not for everyone.
Math understanding is cumulative.	You don't need to remember what you learn. Math doesn't build.
All students are math people when math instruction is taught in accessible and inclusive ways. Math confidence can be built if students are supported to struggle productively.	It's okay if some students don't see themselves as "math people" or believe they can't do math.
Math is a required life skill and is used from furniture placement to budgeting.	Reading is a required life skill. Math is not.
All people, regardless of demographic markers, use and excel at learning mathematics. In fact, the field of mathematics benefits from having more diverse minds engage with it (Goffney et al., 2018).	Some students, especially male presenting students, are more mathematically inclined than other students, like female-presenting students.
Some people have had the privilege and support to engage more frequently, creatively, and flexibly with math.	Some people are just better at math.
Everyone can and does think and reason about mathematical ideas and concepts.	I don't have a math brain.
When people engage in worthwhile math tasks, they will experience productive struggle in learning mathematics.	If you are good at math, it should be fast and easy.
Math is about sensemaking. People who engage in math should understand why it works. This includes application of efficiencies, identities, or shortcuts.	Math is a collection of tips, tricks, and shortcuts.
Doing math with understanding, depth, and thoroughness is doing it well. Speed can be the result of prior experience and practice.	Being fast with math is evidence of doing it well.
Confusion is a necessary step toward a deeper understanding of mathematics. Confusion indicates a restructuring of current knowledge to accommodate new information.	Confusion means you don't know what you are doing.

As you reflect on this table, you may be thinking that despite the myths, you enjoyed math courses as a student. If that's the case, that is great and may be the result of support to engage leisurely with math, an affinity for puzzles or games that use math, or access to people in your life who consider themselves math people. You may even have enjoyed the ways that math allowed you to feel smarter, quicker, or better than those who struggled in your math classes. Our question to you is, what about those who didn't have those supports and didn't enjoy mathematics or the opportunities that understanding mathematics opens up?

In contrast, you may have been a person who didn't feel capable in math classes. If that is you, revisit the faulty beliefs above. Are there ways that you recognize how these beliefs may have contributed to your experiences or the experiences of people you know?

As you ponder these beliefs, you might also be thinking that it is okay if some people experienced mathematics through the lens of faulty beliefs. You think "not everyone needs math to be successful." While that may be true, we advocate for all students to have the opportunity to engage deeply with mathematics. For us this goes beyond being able to choose a specific career, though many careers use mathematics in creative ways, it is about being able to interpret the statistics used in news stories, making sense of financing deals, and understanding how to build wealth or address debt. Ultimately, it is the responsibility of teachers to teach mathematics in ways that allow students to understand, think, and reason as reasoning, critical thinking, and collaboration are life skills. They are components of productive, informed citizenship. These skills need to be a part of mathematics in the same way they are in language arts, science, or civics classes.

HOW DOES THIS BOOK HELP?

This book is based on our years in education, as classroom teachers, math coaches, department chairs, and professional development providers and creators. We have spent our careers as learners, whether that be while taking courses, doing research, or learning in our classrooms every day. The questions this book poses are from our own experience and from what the teachers we work with ask. Our answers are based on our experience, practice, and research.

The book is organized into five categories framed by overarching questions:

1 How do I build a positive math community?
2 How do I structure, organize, and manage my math class?
3 How do I engage my students in math?
4 How do I help my students talk about math?
5 How do I know what my students know and move them forward?

These questions are the Five to Thrive for your secondary math instruction: community, classroom structure, student engagement, facilitating discourse, and using assessment.

Interspersed there are sidebar notes on fostering identity and agency, access, and equity, teaching in flexible settings, and related Great Resources for deeper learning. This book is not an in-depth look at any of our topics. There are fantastic books that cover each of our topics in more depth, many of which are mentioned in the Great Resources sidebars and our list of references. We wrote this book to be easy to read and to provide reliable and practical guidance as you work to improve your craft as a teacher.

WHO IS THIS BOOK FOR?

Ideally, this book is for every educator involved with mathematics education in your building and district. Part of this audience is those new to teaching secondary mathematics, whether they be a beginning teacher or someone whose assignment has changed to include mathematics for the first time. At first glance, that may seem to be all of whom this book is for. But it also is for preservice teachers to use as a way to focus their reflections when they are observing classes and as an aid when they start their own student teaching. That means that the instructors in charge of preservice education and the supervising teachers should at least be aware of what this book offers so they can recommend it and use it as a resource.

This book is also for veteran teachers who want to learn more about implementing current practices in mathematics that they have not had the time to research and consider using in their classrooms. Trying something new to them, perhaps setting common classroom norms, should be quite manageable using this book. Since one of the roles of mentors and mathematics coaches is helping teachers improve their own practice, this book is useful by providing specific guidance on how to implement effective teaching practices for mathematics for the teachers they work with.

Another part of the audience is special education teachers, multilingual teachers, paraeducators, and those who coteach or support mathematics education. Knowing answers to questions about providing differentiation and supports, as well as learning more about building learning communities are valuable assets. Finally, we are hopeful administrators (including district math supervisors) will use this book to deepen their knowledge of best practices for teaching mathematics. The structure of this tool is useful for one-on-one conversations, grade-level meetings, and district-level meetings, and also as a basis for classroom walk-bys and observations because of the focus on best teaching practices.

HOW SHOULD YOU USE THIS BOOK?

One of the quotes you will see as you read the book is from Dr. Juli K. Dixon in which she differentiates between just-in-case intervention (frontloading all the issues you think will occur and trying to address each of them, whether knowledge of them will ever happen to you or not) and just-in-time intervention, meaning providing supports when they are needed and specific to your questions. This book is designed around the latter of these. Use it as a guide for your own questions or when helping others when the need arises. For example, you may have a question about how you will improve communications with caregivers, and there is a question specifically about that. You may have just been observed and asked to change how you are writing your lesson plans, for which there is guidance in this book. Another way to use the book is to do a read through at the beginning of the year so you know what is in the book and some of the big ideas of mathematics pedagogy and are then aware of what topics you have information about and can dip in and out as needed and as your practice grows. Go easy on yourself and know that you don't need to change or master everything right away. Give yourself the grace to grow your practice over time.

Finally, look at the closing section to learn about options you have for pursuing growth in your own learning, and consider how to use the resources at the end of the book, as well. Becoming the best teacher you can be is not an easy task. There will be days when you feel everything went perfectly and times when you make mistakes. Enjoy the good times and learn from the bad times! Your journey will be exciting and fulfilling.

ACKNOWLEDGMENTS

We are grateful to the ideas and contributions of the many leaders, researchers, and teachers in mathematics education that have influenced our understanding of what it means to teach math well. Julia Aguirre, Robert Q. Berry, III, Rochelle Gutierrez, and Danny Martin have influenced our pursuit of equity in mathematics teaching and learning. You will see that works by significant mathematics educators such as Deborah Ball, Jo Boaler, Juli K. Dixon, Barbara Dougherty, Carol Dweck, Ilana Horn, Peter Liljedahl, W. Gary Martin, Margaret (Peg) Smith, and many more have shaped our understanding of math content and pedagogy. You will recognize that Gloria Ladson-Billings and Zaretta Hammond have helped us understand what culturally relevant mathematics should look like in our classrooms.

We have relied on truly exceptional works that support teaching mathematics content. Many are noted in the Great Resources sidebars and the suggested resource list on page 153. For example, the different editions of *5 Practices for Orchestrating Productive Mathematics Discussion* (Smith & Stein, 2011) have been at the top of our list of resources since they were published.

And of course, we have seen some excellent teaching in action, influenced by these research-based resources, but really crafted by teachers as they worked day-by-day to understand what works best for their students in learning math.

chapter ONE

HOW DO I BUILD A MATH COMMUNITY?

This chapter gives guidance for creating the mathematical community you envision. Your classroom should be a community of learners coming together that encourages them to learn from one another in an equitable environment. One aspect of your mathematical community is the mathematics content. You want your students to experience success and to be capable thinkers and doers of mathematics. Students should be positioned as the authors of ideas and as sense makers in the classroom. This chapter points to ways of fostering a positive math identity and a strong sense of mathematical agency in your students that empowers them for success.

The other aspect of the community you envision is how students interact with each other. Your community positions students as contributors to the collective knowledge of the class. Your community has students working together as active participants, who are responsible for explaining their reasoning to their peers and teachers while also making sense of the reasoning of others. The students assume a shared responsibility for their learning as well as the learning of their peers and recognize each other's contributions. Students' strengths are the grounding for these behaviors, which helps let students know they are valued for what they bring to the classroom.

> Some of my best lessons have been ones where students knew how to contribute and felt able to do so. These lessons made it easier for students to buy-in to our class norms.
>
> —MIDDLE SCHOOL TEACHER

Your mathematical community extends beyond your classroom walls as it engages families to partner in the work. We are using the phrase caretakers when we talk about families to reflect the diverse circumstances each student brings to the classroom.

It is very important that your mathematical community is a safe space for each and every student. It is an inclusive environment that values diversity and different cultures. It is a space where norms are set to encourage participation, support of each other, and care for each other. With you as a model, students will work, speak, and act with each other in respectful ways. Student-centered and culturally relevant teaching practices are looked at, as well as equitable mathematics instruction. This chapter helps you create and maintain your classroom community by providing answers to questions including the following:

- ☐ **What is equity in mathematics?**
- ☐ **How do I build and sustain a positive mathematical community in my classroom?**
- ☐ **What norms should I have in my classroom?**
- ☐ **How can my words and actions focus on students' strengths?**
- ☐ **How do I learn about my students' math identities?**
- ☐ **How do I support student agency in my classroom?**
- ☐ **How can I make math class more student-centered and culturally relevant?**
- ☐ **How do I establish two-way communication with caregivers?**

As you read about these, we encourage you to reflect on the following questions:

- ☐ **What does this mean to me?**
- ☐ **What else do I need to know about this?**
- ☐ **What will I do next?**

What Is Equity in Mathematics?

When we talk about equity, we are taking into account all students. We are not talking about any one social status, racial group, gender group, or any type of sorting or demographic data of students. Equity is not code for some traditionally underserved part of the learning community. Equity is about high expectations for each and every student. It is the belief that all students bring knowledge into the math class, that math is for every student, that every student can learn to do math, and that every student deserves an opportunity to learn it meaningfully.

EQUITY IS NOT EQUALITY

Equity is about providing accommodations for students. It is easy to confuse equity with equality, but they are not the same thing. Equity does not mean that every student should receive identical instruction; instead, it demands that reasonable and appropriate accommodations be made as needed to promote access and attainment for all students. An equality model says each student may be given the same, that is equal, support, which may result in some students not having the specific support that they need (see Learning Needs, p. 90). Equity is about ensuring each and every student is mathematically successful by also considering the needs of the most vulnerable in the learning community and providing the supports and accommodations needed (see Learning Needs, p. 90) to make sure all students have the same likelihood of achieving mathematical outcomes.

EQUITABLE TEACHING PRACTICES

With this goal of ensuring each and every student is mathematically successful, we describe equitable teaching as the pedagogy and practices we use to ensure that each student has access to high-quality, on-grade level curriculum and instruction in a learning community that values their contributions.

AN IMPORTANT COMPONENT OF EQUITABLE TEACHING IS HIGH EXPECTATIONS

Sometimes teachers think that lowering their expectations is in the best interest of the student (e.g., when a student is clearly struggling, is missing school frequently, or seems like they are giving up). What this looks like may be deciding not to teach a specific topic ("My students are not ready to solve systems using elimination, so I'll just use graphing and substitution.") or only requiring students to attempt the easiest problems in a problem set. (Please note, this is not the same as accommodations on an Individualized Education Plan (IEP) that address workload as an accommodation but is a teacher-initiated action in response to perceived struggles or deficiencies of students). Another way of reducing expectations is to give hints that take away opportunities for students to productively struggle (see Productive Struggle, p. 81). Rather than doing these moves, consider instead more equitable teaching moves:

- Teach on grade level and use appropriate mini-lessons that address needs or foundational skills as they occur.

- Use a strategy that encourages students to depend on one another such as "Ask 3 (before you ask me)" or intentionally distancing yourself from students in the first few minutes of working on a task.

> *I often set a three-minute timer when I give students a new task. They know that before the timer goes off, they need to get started on the problem using their resources. That could be notes, an elbow partner, or our textbook.*
>
> —HIGH SCHOOL MATH TEACHER

- Ask cognitively challenging problems. That is, selecting cognitively challenging tasks and believing that your students can and will rise to the occasion since they are capable of making sense of and persevering with these challenging tasks (see Tasks, p. 68).

TRACKING IS A STRUCTURE THAT CAN LEAD TO REDUCED STUDENT ACCESS TO QUALITY, ON-GRADE CURRICULUM

For example, students in lower-tracked classes are denied deep learning experiences and relegated to cursory skills work. High-impact teaching practices are for every student. Simply put, don't reserve the "good teaching" for the students that are the most compliant, most excited, or most academically successful. Every student deserves good teaching from teachers who want to teach using equitable and effective teaching practices.

RELATED TO TRACKING IS HOW STUDENTS ARE PUT INTO SMALL GROUPS FOR IN-CLASS WORK

Grouping should not be homogeneous based on some perceived attributes or prior academic achievement, but instead should be random so that students work with different peers and are exposed to different thinking (see Grouping, p. 57). When formative assessment data show that students are lacking a needed skill or conceptual understanding, teachers must find ways to include this in instruction that moves students forward on the current course trajectory. That is, you do not stop current instruction to reteach, but you include the supports while addressing the course standards. For example, when students struggle with fraction computation, that skill can be addressed as part of solving equations, finding area and perimeter, or looking at rates of change.

EQUITY INFORMS EVERYTHING YOU DO

It's especially important to understand that equity isn't an add-on or an extra burden. What if students don't have materials such as a writing utensil (pencil, dry erase marker)? That is an equity issue. While it may not be feasible for you to provide writing utensils for every student every day, it's not equitable for you to deny a student access to the work because they don't have some materials. Providing these items, even just for your class, means that you have removed a barrier to learning for your student. Use of technology is another issue of equity, you must consider student access to technology before assigning outside of school practice that requires the use of technology (see Technology, p. 83).

HOW DO YOU KNOW WHAT YOU ARE DOING IS EQUITABLE?

Consider these questions:

- What evidence do you have that your practice treats all students as capable mathematics learners and doers?
- How might you need to alter your instruction to offer more support to more students? Supports might look like spending more time on a subject, instruction on how to take and use notes, or access to a range of manipulatives or tools to support finding problem solutions.
- Do you know your students' perspectives on this subject? Have you asked them?
- When you are reading this, what students are you thinking about? Who aren't you thinking of? Why?

EQUITABLE MATHEMATICS INSTRUCTION IS ABOUT DEVELOPING A COMMUNITY THAT IS COHESIVE AND COHERENT

Everyone is in this learning process together and what we are doing makes mathematical sense to everyone, including you, as the facilitator. Equitable teaching informs students' identities (see Math Identity, p. 24) in that it helps them see that learning math positions them as capable learners and doers of mathematics that author their own ideas. Students are active learners, so they develop agency in themselves by participating in mathematics in ways that are meaningful, both personally and socially (Berry, 2016). The tasks you choose (see Tasks, p. 68 and Critical Thinkers, p. 76) make student reasoning an important factor of the classroom—everyone's thinking counts and is valued, which supports students in viewing themselves as having ownership of the mathematics to foster identity and agency.

The table below provides a listing of five equity-based instructional practices along with brief descriptions of them (Aguirre et al., 2013). As you continue to form your own teaching style, be sure to include these practices to offer more learning opportunities to your students.

Five equity-based practices in mathematics classrooms
Go deep with mathematics: Develop students' conceptual understanding, procedural fluency, and problem solving and reasoning
Leveraging multiple mathematical competencies: Use students' different mathematical strengths as a resource for learning
Affirm mathematics learners' identities: Promote student participation and value different ways of contributing
Challenge spaces of marginality: Embrace student competencies, diminish status, value multiple mathematical contributions
Draw on multiple resources of knowledge (math, language, culture, family): Tap students' knowledge and experiences as resources for mathematics learning

How Do I Build and Sustain a Positive Mathematical Community in My Classroom?

Think about your students and how you want them to learn. Discussing ideas, collaborating on solutions, feeling free to share their thinking, assisting each other in learning—these are all hallmarks of a productive learning community, as compared with traits such as compliance with rules, students always working only independently, and students being quiet and not sharing thinking. Each teacher has the capacity to build a mathematical community that reflects their beliefs and the identities of their students, which means no two communities will be identical. It is your words and actions that will ensure each and every student feels respected, valued, and important to the class.

HOW DO I CREATE A MATHEMATICAL COMMUNITY?

A classroom community does not form on day 1, nor does it remain static once its beginnings are established. You must make it clear that your intentional actions are fostering a mathematical learning community.

Tip 1 | Plan for what you want your community to look like instead of what you don't want it to look like

For example, **I want my students to discuss and share their ideas** rather than **I don't want my students working quietly on their own.**

Tip 2 | Establish the norm of respect for each other by modeling what you expect

Greet each student by name as they enter the room each day.

Express your genuine interest in getting to know your students, doing things such as learning their names and their correct pronunciations, asking about something they have shared, and so on.

Model inclusive behavior that respects different socioeconomic groups, different races and religions, students with special needs (language, adaptations, etc.), and genders (follow district protocols for using students' pronouns). For example, avoid situations where you compare two genders for a context or where you reinforce stereotypical behaviors (e.g., all boys like football). Intentionally highlight mathematicians from different cultures, races, and orientations in brief discussions, student reports, or wall decorations to honor differences and increase a feeling of being welcome.

Tip 3 | Decorate your walls

Think about what should be on the walls. Post **motivational posters** about thinking, about **positive attributes** such as learning through mistakes, and about **different**

Great Resources

Women You Should Know Downloadable Posters of STEM Women Innovators: https://bit.ly/3lTOs17

Department of Energy Women in STEM Posters: https://bit.ly/3AAWOkl

11 Famous African American Mathematicians You Should Know About: https://bit.ly/3CwxLjc

Math for America Racially Expansive STEM Histories: https://bit.ly/3lVoyKs]

How Do I Build and Sustain a Positive Mathematical Community in My Classroom?

15

mathematicians from a variety of ethnic, gender, and social backgrounds (see Start of School Year, p. 38).

Tip 4 | Arrange your room

If you share a room with other teachers, meet to decide **what works for all of you**. This may mean taking time to move seats before and at the end of class.

If you have a permanently set room (perhaps moored tables), decide how you can work with that.

Tip 5 | Celebrate successes

Display student accomplishments including exemplars and unique solutions. Be aware of sharing from a variety of students. Ask students for permission to post their work before sharing to address confidentiality.

Reinforce positive community moves through compliments such as, "I noticed your group did a great job making sure everyone was part of the solution."

Develop group goals. Keep a running total of checkmarks for each group as it displays behaviors that are part of your class norms. Have a fun reward when a group meets a predetermined number of checkmarks. This reinforces classroom norms and sends a message that everyone is working together.

Tip 6 | Be firm and friendly

Students do not need an adult friend who is "just like them." Students need **someone who facilitates their learning and who ensures norms** for effective learning are maintained.

SUSTAINING YOUR CLASSROOM CULTURE

It takes work on the part of the teacher and the students to nurture and grow a mathematical culture. A key part of this is revisiting and revising the classroom norms the group agreed to during the first week of school (see Establishing Norms, p. 18). This makes the hard work of getting students to work together and to share ideas easier for everyone. If the class is not working from a shared vision of how the class should work, your mathematical and social goals will be harder to meet.

Goal	What to do	What not to do
Each and every student is part of the mathematical thinking of the class.	Establish a routine for solving problems, such as "five minutes alone time, 10 minutes group time, sharing out" (see Routines, p. 55)	Employ an "I do, we do, you do" classroom routine.

Goal	What to do	What not to do
Each and every student is expected to be part of discussions and solving problems.	Select and implement low-floor, high-ceiling tasks that are accessible to everyone (see Tasks, p. 68).	Select a task that has barriers for some students such as language or being too far beyond current learning. Use tasks with only one or limited solution path(s).
Each student is important to the class.	Share different solution paths for a task. Highlight solutions of a variety of students. Make connections between and among solution pathways (see Lesson Planning, p. 51).	Share one solution only. Defer to using a small group of students for the majority of the student thinking that is shared.
Foster and build a growth mindset.	Model your high expectations for all students and your belief that all students can attain your goals.	Give math activities that are easily accomplished or solved, explaining places where you anticipate struggle before students start.
We learn from our mistakes.	Share solutions by pointing out the good thinking in them. Ask, "Why did I choose to share this solution?" so students can look for the learning that grew from an error.	Only share correct answers.
Students are responsible members of a learning community.	Highlight when you see desirable behaviors. "I notice several groups are working as a team to solve this problem." Talk to students who are not following the norms individually with as much privacy as possible (see Start of School Year, p. 38).	Call out students for breaking the norms. Impose classroom punishments when norms aren't followed.

I remember from my own days in school how I hated the cold call. I felt put on the spot and embarrassed. I avoid doing that because I know how negative an effect it has.

—HIGH SCHOOL GEOMETRY TEACHER

How Do I Build and Sustain a Positive Mathematical Community in My Classroom?

What Norms Should I Have in My Classroom?

The first week of school is one of the most exciting times of the year. Hallways are bustling with students coming to school eager to start a new year. New middle school students may be nervous about the transition from their elementary experience. Teachers and counselors stand in the hallways to direct new ninth graders to class. Teachers, new or returning, may also feel those first-week jitters. The first week of school is one of your most significant opportunities to begin creating and shaping your mathematical community. Setting your norms and rules is critical for establishing your new community (see Community, p. 15).

WHAT IS THE DIFFERENCE BETWEEN RULES AND NORMS?

Creating rules and norms is how you establish the mathematical community you want. Rules are explicit and nonnegotiable regulations presented by the teacher and school. For example, schools may have rules around dress code or uniforms. Norms are a bit different. They are collective agreements for how students and teachers will work, speak, and act with each other. An example of a norm is, "Students speaking respectfully to each other and the teacher." Rules and norms are the pillars that you and your students build together to guide your classroom community.

HOW DO WE CO-CONSTRUCT OUR NORMS?

The first step is to ask your students for input! Middle and high school students have years of experience in school. They already know what makes a classroom run smoothly and what makes one disruptive. More importantly, they will feel validated and respected that you asked them. Whatever norms are set are expected of all members of the classroom community, and you and your students will have to hold yourselves to a high level of accountability to practice those norms. Creating norms together will allow students to follow them more genuinely. Facilitating a class discussion with your students about norm-setting should reflect your teaching style and show that you are open and willing to hear their responses. The following are a series of suggested questions and facilitation techniques, which you can modify to fit your style.

- Ask three questions at the beginning of the year to get the ball rolling:
  - *What should teachers do?*
  - *What should students do?*
  - *How should we treat each other? How should we talk and act toward each other?*

- Ask students to respond in writing independently first and then talk to a partner or group; then bring the class together and ask each student to share one norm they wrote down or heard from their group.
- Find a way to write down publicly the class norms students are sharing. One way is to make posters and write the class norms down as you go around the room.

Answers to Your Biggest Questions About Teaching Secondary Math

Another way is to type the norms out and display them for students on a projector. It is okay if a student repeats a norm! Place a checkmark next to any repeated norms.

Teaching in Flexible Settings

If your class is hybrid, consider projecting your posters using an online whiteboard or document like Jamboard (https://jamboard.google.com), Padlet (https://padlet.com), or Google docs (https://www.google.com/docs/about/) to allow virtual and in-person students to participate.

If you choose to make posters, make them ahead of time. Title one poster, "Teachers should . . ." and the other "Students should . . ."

Great Resources

Norms construction protocol https://bit.ly/3zApybQ

Forming ground rules protocol https://bit.ly/3iO56eG

COMMUNITY

- After all the class norms are shared, ask any students if anything is missing from the list. This gives the last opportunity for students to share any class norm that may be important to them. If a student says a class norm that you cannot agree with, be honest and explain why the class will not follow that norm. Propose another class norm that you agree with instead.
- In the end, read each of the class norms aloud for the class, tell students you agree to follow the teacher norms.
- Express to students that you expect them to do their best to follow their norms. Ask your students to review the student norms and also agree to follow them.
- Throughout the year, praise students when they follow a class norm. Celebrate the good things happening in your classroom as much as possible.

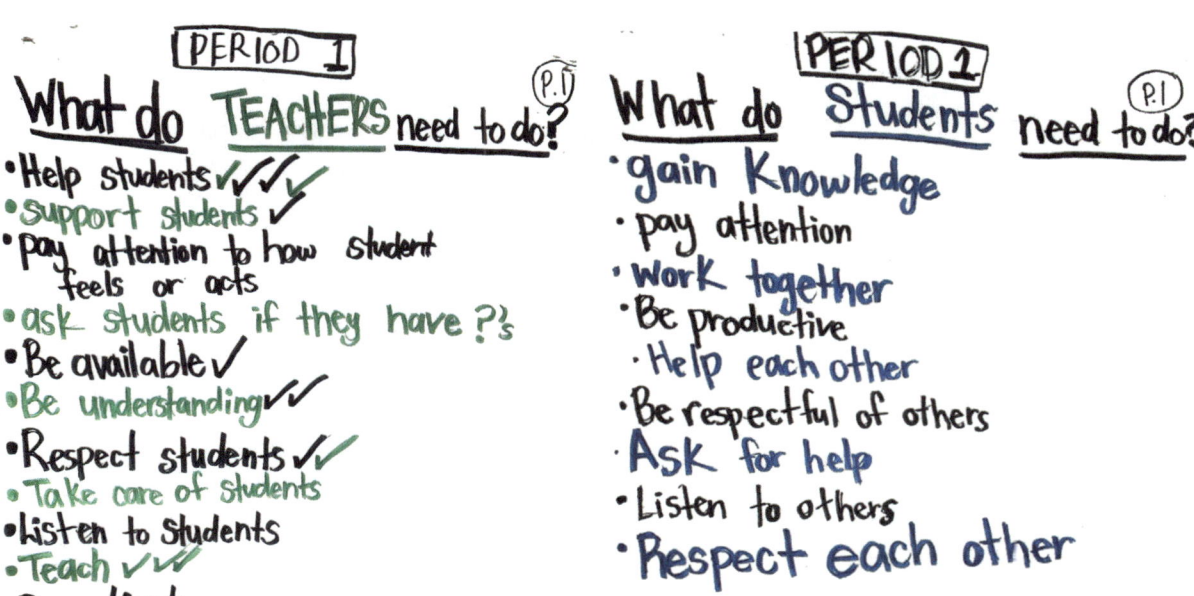

WHAT HAPPENS WHEN A STUDENT BREAKS A NORM?

Think about what you will do when a student breaks a norm and how you want to address the issue. Be thoughtful and restorative in your approach. For example, student A curses at student B. Ask student A to step outside so you can have a conversation privately. Once you and the student are alone, have an honest and open chat. Below are some suggested questions adapted from restorative justice techniques to guide you.

- What happened?
- What were you thinking of at the time?
- What have you thought about since?
- Who has been affected by what you have done?
- In what way have they been affected?
- What do you think you need to do to make things right?

Some students may not be ready to have this conversation. Or perhaps you feel unprepared to facilitate. Consider asking your administrative staff to help. If this happens again, think about how you want to ask for more support from caregivers or administration. Many students act out because there is something else that is bothering them. Make any referrals your student may need to receive the proper services.

WHAT ABOUT RULES?

In addition to your norms, you may have some nonnegotiable rules in your classroom that you want to implement. For example, you could have a "no phones" rule for your classroom. Be sure to rephrase this in the positive by using the phrase, "keep phones off and away." When students break a rule, refer to the restorative justice practices above for some approaches on how to correct the behavior.

Teaching in Flexible Settings

It is important to know your school's rules before writing your own, as you will be expected to hold students accountable to them.

How Can My Words and Actions Focus on Students' Strengths?

Your school year starts with you getting to know your students. You will serve them best if you purposely adopt a positive mindset toward them. Maximizing the strengths your students bring to the classroom (e.g., perseverance or being able to work well in a small group) allows you to help them build on their strengths to become successful mathematics learners and problem solvers.

HOW DO I ADOPT A STRENGTHS-BASED MINDSET?

When you talk with your fellow professionals, you will likely hear two kinds of comments about students. Some educators focus on the positives the students bring each day. They emphasize what students can do, not what they cannot do. Some students will have strong algorithmic skills, while others are good at nonroutine problem solving. An asset approach makes students feel important to your class because each student knows they are contributing something valuable to the whole and knowing their strengths give them points of entry into problem solving. Contrary to an asset mindset is a deficit mindset. A deficit mindset focuses on perceived student shortcomings, such as a student not knowing something the teacher would like them to know or a student being disinterested in class.

These are some phrases to keep in mind as you grow an asset orientation toward your students.

Asset phrases/sentence starters	Deficit phrases/sentence starters
It's good that . . .	These kids can't . . .
That is a unique way to solve that problem	My students didn't learn this last year
All students can learn math, but the time needed and path there will vary	Slow learners
This is a good task for all students because it has multiple access points	Only my high-level students can do these types of problems
I like the way you . . .	Student doesn't know basic facts
My students understand . . . Let's build on that to help them understand . . .	Low ability

Teachers who focus on students' strengths are likely to see greater learning gains than teachers with a deficit mindset. Included among the benefits for teachers who focus on students' strengths are increased student engagement, fewer discipline issues, improved student math identity, and better overall achievement of their learning goals. Building confidence in each student by recognizing and using their strengths helps them to attain your mathematical learning goals and gives them the security that makes them more likely to share their thinking.

HOW DO I UNCOVER MY STUDENTS' STRENGTHS?

Great Resources

Kobett, B. M., & Karp, K. S. (2020). *Strengths-based teaching and learning in mathematics: Five teaching turnarounds for grades K–6.*

Boston College's samples of student surveys and class activities: https://bit.ly/3o1VZOq

The first step to building off of students' strengths is to identify them. What do they think they are good at? What can you observe that are strengths they may not be aware of (e.g., their perseverance, their ability to explain things to their peers, or their positive attitude toward problem solving)? This is subtly different from discovering your students' mathematical identities and is equally important (see Math Identities, p. 24).

A student survey or a writing assignment about student experiences and attitudes is a solid starting point.

Although a survey is easier to use when gathering information, a brief writing assignment can offer deeper insights as it is more open ended. Questions may be as simple as, "The things I am best at in class are . . . ," or "The one thing I wish I could do better in math is . . ." These are prompts that press the students to consider their performance in class. A math autobiography wherein students write about past experiences can also be enlightening.

Choosing a few good problems to start the school year also helps you identify strengths. A problem-solving task that is not based on the curriculum (Liljedahl, 2021) and that has a "low floor, high ceiling" is a good starting point (see Start of School Year, p. 38). You can note strengths such as who communicates their thinking well, who helps their peers to think without giving answers, or who sticks with the problem without giving up. These are assets you will leverage as the school year continues.

USING STUDENTS' STRENGTHS

The following chart is a useful tool for identifying your students' capabilities and deciding how to focus on them when teaching.

Strength	What it means	How to utilize in class
Procedural fluency (This does not mean computes accurately with speed.)	Uses algorithms for a task with flexibility Explains steps of a process with conceptual understanding Likes to write out their work Edits steps out of a process	Have opportunities for students to record their steps and thinking Have tasks that allow students to consider an abstract or symbolic approach, and then ask them to make other representations to solidify their understanding

Strength	What it means	How to utilize in class
Conceptual understanding	Makes connections between different representations relatively easily Wants the mathematics they are doing to make sense	Have opportunities to connect to prior knowledge and other representations Have built-in reflection time so the students can make their own sense of what their work means
Strategic competence	Likes to model mathematical situations Likes to show their mathematical thinking	Allow students to follow their own thinking or to share their thinking within a group Have problems that encourage multiple representations
Adaptive reasoning	Notices patterns in the thinking of others Relates new concepts and processes to previous learning	Look to these students to help summarize class thinking when moving to whole-class discussions Press students to make connections among/ between related concepts
Productive disposition	Sticks to it when solving problems Enjoys the challenge of new problems	Praise their ability to work through confusion and to continue working when struggling Provide opportunities to see math outside the classroom (see Sense of Wonder, p. 96)

Source: Based on Kobett and Karp (2020, pp. 42–45).

Students do not have just one of these strengths but will have more than one and in varying degrees from each other. Being aware of your students' attributes and working to leverage those assets for each student will help them see their successes and to think of themselves as mathematics thinkers and doers.

How Do I Learn About My Students' Math Identities?

Our mathematical identities are formed by the experiences we have learning and doing mathematics. Those experiences shape our beliefs about our ability to succeed at mathematics and what doing mathematics looks like. Culture and perceptions play into our identity as well as interactions with friends, family, and instructors. A student's individual mathematical identity involves how one sees oneself as a doer of and learner of mathematics, as well as how the student views the knowledge, skills, habits, attitudes, beliefs, and relationships they need to develop to be successful mathematics learners (Aguirre et al., 2013).

THINK ABOUT YOUR OWN MATH IDENTITY

Before thinking about your student's math identities, it is beneficial to think about your own math identity. Teachers' math identities influence how they plan lessons, their belief in their students, how they implement their lessons, and how they assess learning. Becoming aware of your own math identity also helps you guide your own professional development. Here are some questions to consider:

- Did math come easily for me as a learner or was it a struggle?
- Do I value making mistakes in math?
- Do I value solving problems in multiple ways?
- Is math playful to me? Do I like to play math games and puzzles?
- What does "success" look like in math class?
- What does it mean to be good at math?
- Who is good at math?

Great Resource

This is a sample student math survey: https://bit.ly/3o1VZOq

THINK ABOUT YOUR STUDENTS' MATH IDENTITIES

Each and every student will come to you with a unique set of dispositions and beliefs about their abilities in math and their understanding of what it means to learn and do math. Race, gender, sexuality, citizenship, and so on are all identity markers that affect how one is viewed with respect to their math ability and in turn affect how people are treated and invited to do math. By understanding and acknowledging your students' math identities, you can better plan for instruction as well as creating a strong, supportive community of learners.

One aspect of identity is self-efficacy. That is, the belief in one's ability to succeed in achieving an outcome or reaching a goal has been shown to be more important than prior knowledge for growth. When students view themselves as capable doers of mathematics and believe in themselves, they often outperform students who have the necessary prior knowledge but do not believe in their ability to overcome difficulties (A. J. Martin & Marsh, 2006; Multon et al., 1991; Skaalvik & Skaalvik, 2004).

You want to gain useful information about your students, and you want your students to gain insight into their own identities and how they see themselves both in relation to math and as part of your mathematical community. Activities that uncover your students' mathematical identities need to be worthwhile and productive for both you and your students. Here are some suggested activities:

1. Name Tents with Feedback (original idea from Sara Van Der Werf @ saravdwerf), usually done the first week of school, can be used to uncover students' attitudes and beliefs about mathematics. Each student makes a name tent with their name on the outside and a space inside to write comments to and answer prompts from the teacher as well as a response from the teacher. You can also ask students to share their pronouns on their name tents.

2. A mathography is similar to an autobiography with a focus on a student's history with math. Have students write about themselves as an individual (hobbies, talents, interests, race, gender, community), themselves as a student (favorite subjects, importance of school in their lives), and themselves specifically as a math student (favorite topics, how you learn best, how you feel about math). This allows you to get to know each student's background.

3. Write a Dear Math Teacher letter sharing experiences that have formed students into the mathematicians they are today. Students can include what they like/dislike about math and good/bad experiences they have had in the past.

4. Students can write a Math Timeline that includes their first math experiences, favorite math experiences, classes they took, math-related moments they specifically remember, and so on.

5. Journal Prompts can be given to students at the beginning of the year to get a baseline as well as throughout the year to show progress. Prompts could include questions about attitudes in math, such as, "Am I good at math?", "Do I enjoy math?", and "Is math useful?"

6. A start-of-the-year survey could ask, "What should I know about you as a learner to help you be successful in this class?"

7. A Math Beliefs Inventory allows both you and your students to reflect on how they view themselves as learners.

FIND WAYS TO HAVE A POSITIVE IMPACT ON STUDENTS' MATH IDENTITIES

Instructional choices that you make and beliefs that you have affect student engagement, support learning, and develop students' identities.

MAKE INSTRUCTIONAL CHOICES THAT ACTIVATE PROBLEM SOLVING

- Ask students to attempt to solve a math problem in multiple ways and honor the variety of solution paths that are found. Students should be the authors of ideas.
- Incorporate explorations and investigations, where students interact with mathematics in a way that allows them to "discover" or experience mathematics.
- Employ challenging, open-ended, and/or nonroutine tasks that offer new solutions or insights that are unexpected for a student.
- Use collaborative learning groups, such as randomly assigned small groups, because participation is an integral part of learning and encourages creativity.

Teaching in Flexible Settings

Use technology, like Google Forms or Desmos, for these activities

Identity and Agency

Continuous/repeated small, meaningful moments of connections are more impactful than a single flash activity at the start of the year.

- Have students explain their strategies so they are seeing other people's approaches to solving a math problem and ways of processing.

BELIEVE IN YOUR STUDENTS

- Believe unconditionally in each person's mathematical capacity. How you think about each student is tied directly to the way you interact with and treat them.
- Include everyone. When math teachers offer full membership, particularly for hesitant learners, students change their attitudes and beliefs about who is good at math and what success in math looks like.
- Maintain a safe and inviting relationship involving trust and belief in your students' brilliance.

REDEFINE MATHEMATICAL SUCCESS

- Vigilantly view student attributes as assets rather than deficits (see Strengths, p. 21)
- Broaden your view of what mathematical competence looks like (posing great questions, working in an organized way, making connections, finding multiple ways to solve problems) and praise that competence when you see it.
- Normalize mistake-making and value mistakes as opportunities to learn.
- Showcase mathematical brilliance from people of varied backgrounds, cultures, and ethnicities.

Great Resources

Desmos Blog: Hamburger, A., Helft, S., & Moynihan, F. (2021, July 14). Rewriting our list of mathematicians. *Desmos.* https://bit.ly/3LLfckw; arbitrarilyclose Mathematician Project: https://bit.ly/3EDMUku

How Do I Support Student Agency in My Classroom?

One of the most important aspects of creating a classroom community is making sure that students feel like co-owners of the learning space rather than simply consumers. In many ways, developing a space where students feel able to direct how they engage in the classroom can result in more shared awareness of what students' learning needs are. Having a plan for how you want your classroom community to look and feel is an important step toward students exercising agency in your classroom (see Community, p. 15).

Below are some tips to guide you in increasing student agency in your classroom.

Tip 1 | Let students choose what work to display on the wall

Oftentimes teachers choose examples of students' work to put on the walls of the classroom. This contributes to a positive mathematics identity and allows students to show others what mathematics they know. This can increase a student's status in the classroom as well. That is, more classmates will see a student as mathematically capable and having worthwhile ideas to contribute. This can lead to that student being invited into group work more easily and listened to when they share their ideas. This tip hopes to expand on the practice of sharing work.

- Dedicate wall space for student work that students want to post.
- Students who post their own work on this wall can add a short message sharing with others what they are proud of with respect to this work. To help with this, your feedback to students should also include what was notable or interesting about their work so that they know your perspective on their successes as well.
- In middle and high school, students may be less enthusiastic about choosing work to post, so this practice will need to be a large part of celebrations of success in the classroom.
- If possible, try to use the walls that students look at the most to display student work.
- Rotate posted work regularly so that all students have a chance to share something they are proud of with their peers.

Tip 2 | Give students ownership of portions of the lesson

It can be challenging to relinquish control of parts of your lessons to your students, but the outcomes can be well worth it. In the article "Never Say Anything a Kid Can Say!" (Reinhart, 2000), the author shares that when teachers are the ones speaking, they are the ones learning. Since we want students to be the ones learning, they need to be the ones speaking. He shifted his teaching strategy from "one who explains things so well that students understand" to "one who gets students to explain things so well that they can be understood" (p. 478). One way to provide students more agency in a class is to give them opportunities to share their reasoning in ways that best serve them. Some options might be as follows:

Whole class share out	• Allow students to use your document camera to show their thinking by displaying their solution pathways or progress. • Allow students to share places where they have gotten stuck in the solution process with the class to support shared learning.
Small group share out	• Allow students to share their thinking in small groups to gain clarity before whole-class conversations.
Pair share	• Allow students to engage in a work swap where only one person sees and critiques their work so that they can make more progress.

Tip 3 | Allow students to choose the topics of projects

Another way to allow students to exercise agency in their classroom is through choice of what tasks they are engaging in. There are many mathematical concepts that are useful in making sense of the world. Projects that students choose to engage in based on their interests can show students how math is connected to what they care about (see Critical Thinking, p. 76, and Culturally Relevant, p. 29).

- Offer a range of topics to students that allow them to make sense of and model data about phenomena they find interesting.
- Frequently share graphical representations of data that may be of importance to your student (e.g., increase in grocery costs or bus fare, reduction of grocery stores in their neighborhood, impact of drought on water levels in their community) (see Culturally Relevant, p. 29).
- Invite students to share about interesting math uses or facts that they encounter outside of class. This could be connected to current events (Olympic trial times) or local social events (winning scores at school volleyball tournaments, the payout in a lottery and what the probability of winning is).

A word of caution: Student agency is meant to be a safe and structured way for you to increase student buy-in for learning. It is not meant to be a license for students to reinforce problematic norms or behaviors in your classroom. There might be times when students ask to engage in a practice that you believe can be harmful to other students in the room. For example, student agency should not be used to restrict working groups—that is, who students will work with or learn from. Agency is also not to be used to allow students to opt out of the learning altogether. Students can choose *how* to engage, not whether to engage. You should ensure that whatever you allow students to do in your classroom follows your classroom norms and school and district codes of conduct.

How Can I Make Math Class More Student-Centered and Culturally Relevant?

Students succeed more when they are at the center of learning because they learn how to be responsible for their achievement. Likewise, valuing students' cultures and identities helps them feel empowered to succeed in math class, their school, and their communities. These tips are designed to make your class more student-centered and culturally relevant to maximize student performance in learning mathematics.

HOW DO I MAKE MY CLASS MORE STUDENT-CENTERED?

A student-centered classroom is a collaborative learning environment that prioritizes students' learning needs. As soon as you walk into the classroom, it is evident the emphasis on instruction and learning has shifted from the teacher to the students. You will find students talking to each other, staying engaged, and learning math from one another. Here are some tips to make your classroom more student-centered.

Tip 1 | Increase student engagement

- **Ask students about their interests**. Create a survey, discussion, or short assignment where students share privately and/or publicly. Look for common interests across your classes and students. Use these interests to help you connect your lessons with students. Support your students in making connections between their interests and math topics.
- **Give students choice**. Provide choice and options for students when it comes to both learning and showing evidence of learning. Liljedahl (2021) demonstrates how increasing student autonomy through allowing them choices builds engagement and thinking for students (see Agency, p. 27).

Tip 2 | Incorporate inquiry-based lessons, projects, or curriculum

- **Include real-world problems in lessons, projects, or units**. For example, in a geometry project, students can estimate the height of trees on your school's campus using similar triangles and shadows (see Critical Thinking, p. 76, for more examples).
- **Connect real-world events**. Take time during class to discuss current events, especially how they relate to mathematics. There are examples in the media every day that use mathematics, especially statistics. Find examples that align with your student's interests and share when possible.
- **Involve your school's or students' communities**. Showcase student projects or student work during a school event. Ask students to present their work to their caregivers or students from other classes.

> **Great Resources**
>
> *USA Today*, Skew the Math, and *The New York Times* "What's Going on in This Graph" all have lessons built on current events.

Support students as leaders

- **Give students roles in leading class.** Ask students to lead parts of the lesson or activity so that they have ownership of their learning. Rotate students through each role daily or weekly so that all your students have an opportunity to be class leaders.
- **Provide opportunities for students to teach each other.** Students can present solutions to problems to the group or the class. Have students work on shared whiteboards or create posters to display their solutions. Deflect student questions to other students so that students become resources for each other.

HOW CAN I MAKE MATH CULTURALLY RELEVANT AND CULTURALLY RESPONSIVE FOR MY STUDENTS?

Culturally relevant and culturally responsive teaching provides additional benefits for academic success. Culturally *relevant* teaching recognizes the importance of incorporating students' cultural experiences, language, and lived experiences in their learning (Ladson-Billings, 1994). Culturally *responsive* teaching is empowering, transformative, emancipatory, and liberating (Gay, 2000) and provides some of the best learning conditions for all students (Hollins, 1996). Culturally relevant teaching is most *responsive* when it is adjusted to reflect the experiences and cultures of the students in the room. Because your students change year to year, the connections you make to the curriculum must change as well. By sharing different viewpoints or perspectives based on their own cultural and social experiences, students become more active participants in their learning (Nieto, 1996). Here are some ways to strive toward a culturally responsive and relevant math classroom.

Tip 1 Learn about students' culture and experiences

- **Talk to other teachers and staff from the same cultural backgrounds as your students.** Ask them effective ways to connect with students and get tips and advice for student success. Visit or observe their classrooms to learn more about how to make diverse cultures shine or do research on your own.
- **Visit events and communities relevant to your students' culture.** Be open and respectful to different beliefs and customs. Use the knowledge you gained from this experience to improve your teaching and connect with your students.
- **Talk with caregivers.** Ask them about their culture and write down any advice they give you about their student. Be honest about what you can do to support them in school (see Communication, p. 32).
- **Empower students to be critically conscious.** Ask students to investigate systems of power and issues of justice in our society. Find ways for students to actively participate in helping their communities embrace diversity.

Tip 2 Encourage and embrace culture in your class

- **Encourage students to talk about their culture.** In the classroom, advocate for students sharing about their culture (especially as it applies to mathematical thinking, e.g., talking about different algorithms for solving right triangle problems) and take time to listen to what they have to say. Value their cultural experiences and share your own.

- **Discuss stereotypes with students if they come up**. Make students think critically about their generalizations. Have honest conversations with the class about stereotypes and how they can harm our community. Remind students about the norms that they built at the beginning of the year and that diversity deserves to be celebrated.
- **Teach students about mathematicians from different backgrounds**. Teach or ask your students to research famous mathematicians from diverse backgrounds and publicly display them using posters or quotes.

Tip 3 | Celebrate student success

- **Set rigorous and realistic goals for students**. Hold high standards for your students and provide opportunities for your students to achieve those goals. Find ways to display and vocalize their success.
- **Seek and find diverse ways for students to succeed**. Just as every student has different approaches to learning, they also have different ways to show their brilliance. Take the time to investigate how each of your students shines. As time progresses, you'll soon be able to find more ways to celebrate more students (see Strengths, p. 21).

Great Resources

American Mathematical Society mails free posters of mathematicians from different backgrounds: https://bit.ly/3zLXL8B

COMMUNITY

How Do I Establish Two-Way Communication With Caregivers?

Teaching in Flexible Settings

If you have a caseload of 150+ students, consider writing a letter to the caregivers to save some time. Push yourself to make calls throughout the year.

Communicating with caregivers is an essential responsibility for educators, as a way to build a relationship based on trust and understanding. Communicating with caregivers is as much about them communicating with you as you are communicating with them. A good home school partnership will allow both you and the caregiver to best serve the needs of the student by better equipping both parties.

HOW DO I START THE COMMUNICATION?

One way to begin building two-way communication between the caregivers and the teacher is to start the school year with a call home for each student. This allows you to introduce yourself and learn more about the student. In your call, share information about what is going on in class, plans for the year, and ways for the caregivers to contact you. Later in the year, contact caregivers to discuss student progress, positive things you have noted, discipline, or other issues in class. Remember that some parents did not have positive experiences in school so you want your communication with them to be a positive, inviting place to collaborate.

HOW DO I ENSURE A PRODUCTIVE CONVERSATION?

Here are some tips for success:

 *Tip 1* Be kind, open, and gracious

- Enter the conversation believing that every caregiver loves their student more than anything.
- Be cognizant of the caregiver's feelings.
- Bear in mind you are a professional with curricular and pedagogical background who is communicating to help the student, but be careful about thinking (or portraying) that you know more than the caregivers about their student.

 *Tip 2* Communicate to learn

Ask questions to learn more about the student. Consider asking the following:

- When has your child been most successful?
- What are their strengths?
- What has worked well in the past?
- What are their greatest struggles?
- What hasn't worked well in the past?

Tip 3 | Communicate proactively, consistently, and promptly

- Start with something positive to share about the student.
- Once you have established a relationship, learn how caregivers prefer to communicate.
- They might prefer text messages, emails, or voice messaging apps. Seek out tools that don't require synchronous availability.
- Return calls and/or emails within 24 hours.

Tip 4 | Communicate respectfully

- In any communication, imagine both the caregiver and the student are present and never say something to one you wouldn't say in front of the other.
- Your words matter; always take care to make sure they are kind and uplifting.
- If you're feeling angry, be sure to take a cooling-off period before making a call or sending an email. If you do write an email while you are still worked up, don't send it right away. Going back and revising the email before sending will give you the opportunity to soften your language and make sure your communication is clear, positive, and helpful.

Tip 5 | Communicate early

- Calling with good news during the first few weeks of school will set the tone for the relationship, so if something comes up that is more difficult to communicate, the caregiver is wil ing to listen.
- Call 2–3 caregivers a night the first few weeks of school and in no time your whole roster will be covered. You may be surprised how good you feel after telling a caregiver how awesome their child is.
- Start with something like, "let me tell you about one of my favorite things that happened today" and be very clear and concise about what their student did that you liked.

Tip 6 | Communicate often

- Keep notes of your calls/communications for each student that detail the reason for the call, what was discussed on the call, and the agreed-on next steps.
- Review your notes before making additional contacts.
- There are times when you won't be able to accommodate a parent's request. Instead say, "Here is what I *can* do for you . . ."

Tip 7 | Communicate at Open House

- Don't assume a caregiver's absence indicates disinterest. In this age of technology, think about how you can make this meeting accessible to those who cannot physically attend.
- Open House is your chance to show your strengths. Be high energy, be encouraging, and love your craft. Your time should be spent on making sure that caregivers know your philosophy of education and having them feel comfortable with you as their child's teacher. Open House is an opportunity to sell yourself!

- Caregivers don't need all the details of your class. They want to know that you have their students' best interests at heart. You can write the details of the class in a syllabus or parent letter. If there are specific details you want them to know, think about the one or two that are most important and share those.

Tip 8 | Communicate at conferences

- Make sure to stay positive and have an "I'm here to help" attitude. An example of a statement that can open up communication is, "Your student is a good communicator, and is very social. How can we focus this strength during instruction?"
- You may need to deflect (or make it into a positive) when a parent is accusatory and/or defensive. Share with parents what you think your responsibility is as well as shared responsibilities. Feel free to say you messed up and take responsibility, then figure out what you can do better next time. Caregivers will appreciate your honesty.

Tip 9 | Communicate at community events

- Consider hosting a family math night once a semester. There you can meet other people who support your students and foster connections between families.
- Invite people who use math in their careers to speak with your students. They could be caregivers or community members. This is another way to connect with the community and with your students.

Notes

chapter TWO

HOW DO I STRUCTURE, ORGANIZE, AND MANAGE MY MATH CLASS?

Creating a mathematical community is not the only thing that contributes to effective mathematics instruction. The ways you structure, organize, and manage your classes—including how you group students and how you plan for instruction—are also important components of effective instruction. How you start your school year and organize your classroom sets the tone for everything that follows. You should be thinking about what learning goals your curriculum outlines as well as the trajectory the goals take over the course of the year. Part of that is knowing what a unit plan is and how it differs from your daily lesson planning. Before you start writing individual lessons, you need to decide what your goals for your students are, both for content and affective outcomes (e.g., building positive identity, how students work together). Then, you turn to your daily classroom practices. The use of the word *routines* for classroom practices means that you have set procedures for how processes and behaviors in class are managed. Another aspect is how you organize your classroom, including how and where students work. Do you want them working on nonpermanent vertical surfaces, or do you want students sitting in groups? You also need to decide how you place your students into groups (e.g., let them choose or randomly assign) and why you opted for that method. Additionally, you need to ensure that your students know the ways you expect them to work together. Finally, decide how you will employ practice and homework with your classes. Building familiar routines helps students know what to do, why they are doing it, and how they should be working. Your structures provide a consistent and predictable learning environment that makes your students feel safe to participate and to share their thinking. Clear and consistent routines applied with necessary flexibility increase student productivity toward your goals.

This chapter answers questions about how to structure, organize, and manage your math class, including the following:

- [ ] **What do I do at the start of the school year?**
- [ ] **What are the process standards, and how do I use them?**
- [ ] **What is a learning goal, and how do I write one?**
- [ ] **How do I plan a unit?**
- [ ] **What makes a good lesson plan?**
- [ ] **What are mathematical teaching routines, and how do I use them?**
- [ ] **What are the ways in which I can group students?**
- [ ] **How should my classroom be organized to maximize student learning?**
- [ ] **What is the role of practice and homework?**

As you read about these, we encourage you to reflect on the following questions:

- [ ] **What does this mean to me?**
- [ ] **What else do I need to know about this?**
- [ ] **What will I do next?**

What Do I Do at the Start of the School Year?

The beginning of the school year sets the tone for everything that follows. Your teacher moves and activities help establish your expectations for your students, set up several of your key routines, and make the initial steps in creating your classroom culture. Additionally, how your classroom is set up and decorated communicates how you view what teaching math looks like, how students will learn, and how students will be treated. Here are some ideas to start the school year, but note that you can implement them at any time.

WHAT SHOULD I DO BEFORE THE SCHOOL YEAR STARTS?

Before you start, think about what your curriculum includes and what you want your classroom culture to be. The former of these means you need to look at district curriculum guides, the materials your district/school have available to you, and the information you can find about past strengths and areas of concern for your students. This can help you think about the trajectories you want your lessons to take, find tasks that fit the cultural and social backgrounds of your students, and plan out a possible timeline for the year (keeping in mind that this will change as you get to know your students better and gain a deeper understanding of their strengths and struggles). You also want to start thinking about how you will build a positive and inviting culture for your students and their families. Part of that includes initiating communication with caregivers before school starts (see Communication, p. 32). Get to know more about your students by reading over IEPs/504 plans and by talking to intervention specialists to learn about strategies you may want to try.

WHAT SHOULD MY ROOM LOOK LIKE?

An important factor to consider is whether you will have your own room or whether you will be traveling among different rooms. If you do not have your own room, you will be restricted in some of what you can do. However, there are several things you should decide.

HOW WILL I ARRANGE MY FURNITURE?

Think about whether you want rows of desks or small groups of desks or whether students will be seated at tables or there will be other arrangements (see Classroom Organization, p. 60). There is a growing movement of teachers subscribing to the classrooms described in Liljedahl's (2021) *Building Thinking Classrooms*, in which there are tables or groups of desks for students to use for storage of materials or individual work as needed as well as nonpermanent surfaces (whiteboards, chart paper, chalkboards) for students to use. In this type of arrangement, vertical, nonpermanent surfaces are used for student workstations. Think about the message you want to convey with your furniture. Does it let students know you value student thinking, participation, and collaboration?

WHAT SHOULD I HANG ON MY WALLS?

To create a positive, inclusive, and student-centered classroom, hanging posters or wall hangings about the value of perseverance, your belief that every student is capable of learning, and motivation are a good start. Also consider the benefit of having displays about famous mathematicians that include a variety of different races, genders, and ethnicities. You may also want to include posters that recognize the diverse cultural wealth in your classroom, especially inclusive and affirming posters. Bulletin boards can have interesting math facts, math recreations (math jokes, logic puzzles), connections to the community, or opportunities for students to share about themselves (voluntarily, as some may not want to partake at the start of the year). For example, students place a card with their name in a Venn diagram with sections such as "I learn from my mistakes," "I am a hard worker," and "I am good at group work"; this sets an asset-based tone and helps you learn a little about your students' identities (see Culturally Relevant, p. 29, and Community, p. 15).

WHAT HAPPENS IN MY FIRST CLASSES?

The first days of school are prime times to pique students' interest. Starting the first day with reading your class rules, going over your syllabus, or distributing materials are not inherently interesting and can be covered later. Here are some tips about things you want to include and address as your year starts.

Tip 1 | Get to know your students

You may want to do a student survey on their beliefs about their mathematical identities (see Identity, p. 24) or their feelings about mathematics. Having students write a mathematical autobiography is another way to learn about your students and their attitudes toward math classes. Use a student strengths–based survey to learn more about your students as well (see Strengths, p. 21).

Tip 2 | Set some of your classroom routines (see Routines, p. 55)

Consider having a noncurricular mathematics task to start the year. Use your routine of randomly assigning groups, so students start to become accustomed to it (see Grouping, p. 57). Establish how you will take attendance, assign tasks, observe student work, offer assistance to students, and go about other procedural routines. As the week progresses, think about using routines such as *Notice and Wonder* (in which students look at a task or set of data and then share what they notice about it and what they wonder about it) or how work is shared in class, which will give you some formative assessment opportunities for learning what your students know (see Routines, p. 55).

Tip 3 | Set your classroom norms (see Collaborative Norms, p. 18)

This is the perfect time to let students know you respect them and want their input. Working together to create your norms gives them voice and creates more student buy-in. After you have done a group problem, ask students to think about what made their work easier to accomplish and what impeded their work. These ideas will be part of setting your norms.

Tip 4 Use engaging tasks

After starting the year with a few nonroutine and highly engaging problems to set your expectations, use tasks that are part of the curriculum that will give you information about where the students are in their learning and help you decide what your next steps are for the class (see Formative Assessment, p. 128). Continue using random grouping, small-group and whole-class discussions, and asset-based language as you build positive student identities and your classroom culture.

Notes

What Are the Process Standards, and How Do I Use Them?

When you are planning lessons, you need to consider what content you want students to learn and how you want them to engage with that content. The *how* of a lesson plan should focus on providing student opportunities to engage as doers of mathematics. While there are many forms of process standards, the Common Core Mathematical Standards use eight Practice Standards that are useful in clarifying behaviors and reasoning skills that are reflective of what mathematics *doers* do (NCTM, 2014). In addition, most districts use these standards or similar ones, so this is useful information regardless of where you teach.

The mathematical practices can be divided into two categories. The first four practices can be seen as behaviors that successful mathematics learners engage in. The final four are tools that successful mathematics learners employ when engaging in math tasks. The table below provides a brief description of each practice and an example of what it might look like if a student were to engage in the practice.

Practice 1: Make sense of problems and persevere in solving them	*Description*: Students approach problems with confidence. They consider what they know about the problem, what it is asking, and how it might connect to prior work. They also keep track of missteps to figure out incorrect solution pathways.
	Example: When presented with a problem, students might ask themselves the following questions:
	What do I know about this? What seems confusing? How is this like other problems I've seen? How is it different?
	While working toward a solution, students don't quit. They try many strategies and can communicate any source of confusion. Throughout problem solving, students ensure that their work makes sense.
Practice 2: Reason abstractly and quantitatively	*Description*: Students know how to strip the context from problems to work with numbers and can refer to the context in order to check their approach. They can develop equations to reflect their thinking. They also share problem solutions with units and in ways that connect to the problem assigned.
	Example: If students received a problem such as, " It takes 1 hour to fill 3 buckets with molasses. How long will it take to fill 8 buckets? Explain how you know you are correct?" They can use the context to set up the direct proportions problem, e.g., $$\left(\frac{1 \text{ hour}}{3 \text{ buckets}} = \frac{x \text{ hours}}{8 \text{ buckets}}\right).$$ Quantitative reasoning might lead to a student writing the following simplified version of that proportion, $\frac{20 \text{ minutes}}{1 \text{ bucket}} = \frac{x \text{ minutes}}{8 \text{ buckets}}$ and finding 160 minutes. The solution they share would be 2 hours and 40 minutes. Finally, the fact that they could create two equivalent fractions could be taken as evidence of their correct solution path (e.g., $\frac{60 \text{ minutes}}{3 \text{ buckets}} = \frac{160 \text{ minutes}}{8 \text{ buckets}}$. They can argue that the solution pathway is correct because the constant for each fraction is 20 minutes per bucket).

(Continued)

Practice 3: Construct viable arguments and critique the reasoning of others	*Description*: Students can justify their strategies and solution pathways. They can make sense of others' justifications and figure out how other solutions are similar to or different from their own. They can develop and communicate the boundaries for when their solutions are true in systematic ways. Finally, they ask questions to learn more about the solution paths that are shared with them.
	Example: For the problem $y = \left(\dfrac{2}{3x}\right)$, students can consider the solution for when $x < 0$, $x = 0$, and $x > 0$. They can use graphical representations to determine the domain of this equation and communicate why it makes sense that $x = 0$ and $y = 0$ are asymptotes for this problem. 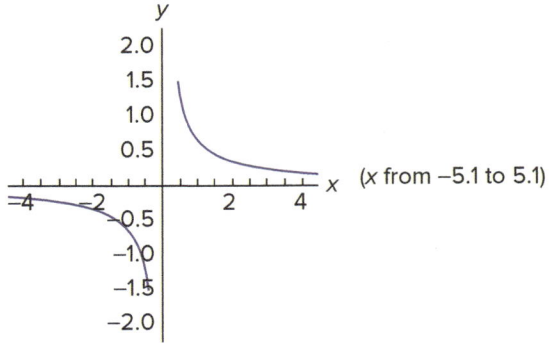 *(x from −5.1 to 5.1)* A student who solved this problem in the form $3xy = 2$ would be able to describe how this form supports their conclusion that neither x nor y can equal zero for this case as well.
Practice 4: Model with mathematics	*Description*: Students can use the mathematics they know to solve problems that they might encounter outside of the classroom. This can look like simplifying a problem to apply a graphical representation to it, developing a diagram to make sense of a situation, or writing an equation or formula to connect different variables in a problem. After simplifying the problem, students can communicate their solutions with the context that they previously stripped away to improve the model.
	Example: For example, you might ask students to share a conjecture for the question: How many golf balls would it take to fill up our classroom? Once students get the dimensions of the classroom and a golf ball, they begin to engage in modeling. That could look like this: • Describing the classroom as a rectangular prism • Using a small cube rather than a small sphere to define the dimensions of the golf ball • Estimating how many additional golf balls could fit around or under furniture using classroom objects for measurement Once students develop an estimate, they can share their conjecture and the assumptions and limitations of their model, and any additional decisions that affected their solution.

Great Resources

See Fermi problems for estimation tasks that support modeling and sensemaking using mathematics; Taggart, G. L., Adams, P. E., Eltze, E., Heinrichs, J., Hohmann, J., & Hickman, K. (2007). Fermi questions. *Mathematics Teaching in the Middle School, 13*(3), 164–167.

The next four practices reflect tools that successful learners employ while solving math problems.

Practice 5: Use appropriate tools strategically	*Description*: Students have access to many tools in the mathematics classroom, including technology (e.g., calculators, computer apps) and manipulatives (e.g., algebra tiles, geoboards). This practice discusses knowing how to use the tools at their disposal, when specific tools could be useful, and which tool is best suited for a task.
	Example: To find the roots of the equation ($y = x^2 + 6x - 4$), students would know that solving the equation $0 = x^2 + 6x - 4$ or graphing the equation $y = x^2 + 6x - 4$ would aid them in finding the roots. They might then decide to solve for x using the equation $0 = x^2 + 6x - 4$ because this strategy and the quadratic formula are something that makes more sense to them.
Practice 6: Attend to precision	*Description*: Students can clearly communicate their thinking with precise language and the appropriate use of labels and units. They also know when they need to provide exact numbers (e.g., using scientific notation) and when estimation will provide an accurate enough response (e.g., $\pi = 3.14$).
	Example: Students label graph axes with awareness of which variable is independent (e.g., time) and which is dependent (e.g., height of a person). And when students then describe the rate of change of the graph, they say it is a ratio of height to time.
Practice 7: Look for and make use of structure	*Description*: Students use past mathematical experiences to expand their understanding of new math experiences. This cumulative view of mathematics supports students in developing their own rules about how numbers relate.
	Example: To support students in making use of structure, share problems that make the structure evident and ask the student what the rule is. Here's an example: In the table below, describe what patterns you see. Find the fractional representation of 3°. Justify your reasoning. <table><tr><th>Exponent</th><th>Fraction</th></tr><tr><td>3^{-2}</td><td>$\frac{1}{9}$</td></tr><tr><td>3^{-1}</td><td>$\frac{1}{3}$</td></tr><tr><td>$3°$</td><td>?</td></tr><tr><td>3^1</td><td>$\frac{3}{1}$</td></tr><tr><td>3^2</td><td>$\frac{9}{1}$</td></tr></table> Students can make use of the table and the fact that the fractions are reducing by one-third each time to find 1 as the solution to 3°.

(Continued)

(Continued)

Practice 8: Look for and express regularity in repeated reasoning	Description: This practice reflects a learner's ability to see when they are engaged in repeated reasoning to make sense of their work. This can commonly be seen in how students notice repeating decimals ($\frac{1}{3}$ = 0.3333 . . .) or simplify the expanded form of a difference of squares.
	Example: If $(x + 2)(x - 2) = x^2 - 2x + 2x - 4$, and $(3x + 3)(3x - 3) = 9x^2 - 9x + 9x - 9$, then any difference of squares will be of the form $(a + b)(a - b) = a^2 - b^2$.

It's important to know the following:

- You cannot teach the practices authentically if the mathematical tasks that students are engaging in are not worthwhile. Students don't need to persevere through procedural problems. Students will struggle to find structure when problems are overscaffolded to remove deeper thinking for them.
- The practices are not something that can be isolated from one another. They work together and are a part of the work of learning mathematics.
- For students to internalize the practices, your learning goals and lesson plans need to include ways you think students will engage in the practices. Anticipating what their engagement looks like will support you in teaching in a way that makes room for each practice. It can also inform the tasks you choose to support students' use of the practices (see Learning Goals, p. 45, and Choosing a Task, p. 68).
- During class, call attention to the ways you notice students engaging in the practices. This kind of targeted feedback can feel more authentic than a general "great work" to students and serves to contribute to the forming of a positive mathematical identity (see Math Identity, p. 24).

Notes

What Is a Learning Goal, and How Do I Write One?

A learning goal or learning target is a detailed statement of the content a teacher plans to teach during a lesson and what it looks like when students successfully learn it. Learning goals should be composed of mathematical content goals—connected to the content standards for your state—and mathematical process goals—describing what mathematical behaviors, practices, or habits of mind students should engage in.

When drafting content-focused goals, it can be helpful to answer the following questions:

1 What will I look for in student work as evidence of an understanding of the concept(s)?
2 What are the various ways students might demonstrate understanding? Is there a difference in sophistication of the strategies I am looking for?
3 How might the math my students already know influence what they produce?

When drafting process-focused goals, it can be helpful to answer the questions below:

1 What will I look for as evidence of students making sense of the problem?
2 What tools might students use to support their reasoning?
3 In what ways might students show or share their reasoning?

Your answers to these questions need to be specific and measurable, so that they can inform your lesson plan's launch, task, and formative assessments. They should also be written in a way that students can figure out whether and how they are making progress toward the goal. Below is an example of a specific, measurable learning goal and one that is too vague to support lesson planning.

Specific	Vague
Content goal: Students will • use the operations of subtraction and addition to simplify equations with one variable, • make use of the structure of inverse operations to find the value of the variable, • demonstrate their understanding of equivalent expressions, and • explain what their solutions mean (e.g., "For the equation $x + 5 = 3$, x must equal -2 for the equation to be true").	Students will learn how to simplify equations with one variable

The *specific* goal is purposely detailed. There is clarity about what students will learn and how they will demonstrate that learning. The goal also points to questions you might ask during a lesson to check students' understanding. For example, a teacher might ask a student to show instances of equivalence in their work to check that students understand the importance of equivalence when solving for a variable. This is the foundational understanding for balancing an equation—something that can be more challenging to keep track of when solving for variables in equations requiring substitutions or other simplifying steps to solve:

(e.g., $5 + \frac{1}{x-3} = \frac{1}{x+2}$).

Finally, learning goals are based on the essential questions, big ideas, and course standards from the unit you are teaching. In the example below, you can see how learning and process goals connect to the content standard, big ideas, and the essential question.

A common eighth-grade *content standard* for expressions and equations is the following: Find solutions for linear equations in one variable. Problems that fit under this standard can be of the form x + 5 = 3.

Some *big ideas* connected to this standard are equivalence and the inverse relationships of operations.

One *essential question* that connects what students are learning about expressions and equations is, "How do we maintain equivalence when simplifying expressions or solving simple equations?"

The *learning goals* for a lesson on solving linear equations in one variable may come in the form of content goals and process goals, such as the following:

- *Content goal*: Students will use the operations of subtraction and addition to simplify equations with one variable. They will make use of the structure of inverse operations to find the value of the variable. They will demonstrate their understanding of equivalence between the different formats of the equation by using the equal to sign to show the relationship between forms of the equation that are being operated on.
- *Mathematical process goals*:
  - Students will *make sense of the problem* by identifying the variable that needs to be isolated.
  - Students will apply their knowledge of inverse operations when working with numerical equations to these linear equations and describe how these equations are similar to those they are familiar with—evidence of an ability to communicate their mathematical knowledge and see connections among the skills and concepts they are learning.

FREQUENTLY ASKED QUESTIONS ABOUT LEARNING GOALS

Q: *Do I share learning goals with students?*

Yes. It is good practice to share a learning goal with students because those who know what the learning target is are better able to assess their own understanding. How and when this is shared is based on a teacher's preference, but much like when students engage in proofs, knowing the end goal should not rob them of opportunities of deep reasoning when worthwhile tasks are employed.

Q: *Won't sharing the learning goals with students give away the "aha" moment?*

No, because students need a student-appropriate learning goal that connects what they have learned in mathematics previously with what they will learn in the upcoming lesson. They do not need the specific learning goal that guides lesson planning, task selection, and formative assessment. A student version of the learning goal above might be, "Today you will expand your application of simple operations to solve for x. We will also continue our discussions of equivalence to continue to differentiate it from equality."

Notes

How Do I Plan a Unit?

Great Resources

For insights on selecting curriculum, see Stein, M. K. (2007). *Selecting the right curriculum research brief.* NCTM. https://bit.ly/3CKcin3

Planning for student learning is one of the most important and challenging aspects of teaching, especially for new teachers. For mathematical content learning goals, there are three levels of planning that affect student learning of the content, all of which are informed by the mathematical content standards. These are the state standards that shape what mathematics students are expected to know and be able to do on completing a grade or course (for more on content standards, see Learning Goals, p. 45). With the content standards as your foundation, the three levels you need to plan for are the following:

- *Curricular plans*: reflect the course standards by connecting them to activities and a pedagogical approach to teaching for the course
- *Unit plans*: plans that connect to chunks of the curriculum that are connected by big ideas
- *Lesson plans*: activity-based plans that support student understanding of skills, concepts, and ideas that relate to a big idea

This section focuses on curricular and unit planning. For more on lesson planning, see Lesson Planning, p. 51.

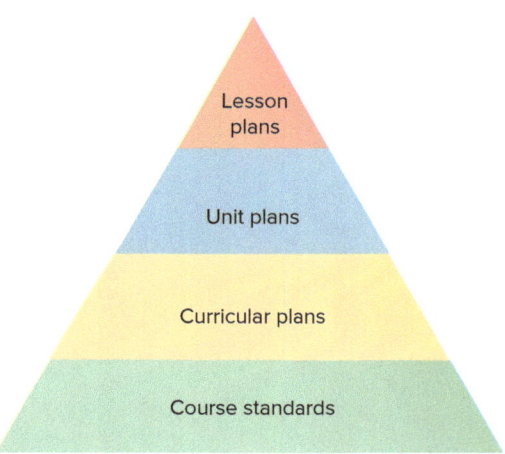

Most districts or schools provide teachers with a curriculum, which might include textbooks, a pacing guide that contains the unit breakdown of the course, and common assessments, among other materials.

WHAT EXACTLY IS THE CURRICULUM?

A curriculum is a holistic view of a mathematics course, including activities and a pedagogical stance that aligns with the course standards; it is much more than a textbook (Stein, 2007). It presents both the trajectory of learning in a course and guides how a teacher might support students in navigating the trajectory. The curriculum is what is taught and how it's taught. Students develop a sense of what math is from activities in the curriculum, and they develop ideas about what it means to do and be successful at math from the pedagogical stance. While all these ideas may not be explicit, the ways you plan and connect learning experiences tell students a story of how norms, concepts, and practices work together to support learning.

HOW DO I MAKE SENSE OF THE CURRICULUM?

Most new teachers aren't expected to plan their own curriculum, but all teachers are expected to make sense of the curriculum that they receive. Here are some brief tips for making sense of the curricular resources that you are given.

- Become very familiar with the content standards for your grade or course.
- Review the pacing guide or curriculum map:

  - This guide may not follow the textbook you are using sequentially. This chronological guide is organized by concept, which may not be the same as the order of the textbook chapters.
  - Note the skills, concepts, and practices that are highlighted in the pacing guide, and make sure you know what standards they support.
- Review the essential question or big idea that drives and unifies the lessons within a unit and understand how each content standard connects to that essential question or big idea. Here is an example:

One of the eighth-grade *content standards* from the Common Core State Standards for Mathematics for expressions and equations is, "Solve linear equations in one variable" (National Governors Association Center for Best Practices [NGACBP] & Council of Chief State School Officers [CCSSO], 2010). Other states have similar standards for eighth-grade math as well. Problems that fit under this standard can be of the form $x + 5 = 3$.

Some *big ideas* connected to this standard are equivalence and the inverse relationships of operations.

One *essential question* that connects what students are learning about expressions and equations is, "How do we maintain equivalence when simplifying expressions or solving simple equations?"

HOW DO I ENGAGE IN UNIT PLANNING?

When engaging in unit planning, it is helpful to work with other colleagues—especially those who have deeply studied the flow and direction of the topics in the units. These colleagues can share insights about where the curricular materials do a good job of mapping onto the standards and when it might be necessary to supplement the curriculum.

- Think about the goals and the flow of the unit. Some useful questions to answer are the following:
  - What should students understand or know how to do by the end of the unit? How will you assess this new understanding? What formative and summative assessments will you use?

Great Resources

See *Fostering Algebraic Thinking* by Mark Driscoll (1999) for insights on big ideas, and *The Mathematics Lesson-Planning Handbook, Grades 6–8,* by Lois A. Williams, Beth McCord Kobett, and Ruth Harbin Miles (2019) for more on essential questions.

- What tasks or activities will you use to support this understanding?
- What lessons, assessments, and/or units should come before and after this unit to support student understanding?
- Decide what combination of lessons will best support students in developing an understanding of the essential questions. This group of lessons is the unit.
- Make sure that there are lessons that support student engagement with each skill, concept, and practice in the unit.
- Review or revise all summative assessments (e.g., any quizzes, tests, or projects) to make sure the assessments reflect what is being taught in the lessons and the essential question students will be working to understand.
- Skim the textbook or other curricular resources to find quality tasks that connect with the standards, big ideas, and skills you have recorded.
- Jot down a brief story of the unit. Think about how you want to move from the essential questions to the final summative assessment. This short story can help you figure out if there are any holes in your unit plan. It also helps you explicitly record how students will address the mathematical practices in a unit.

Notes

What Makes a Good Lesson Plan?

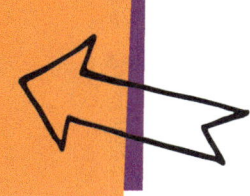

Good teaching does not happen by chance. Good teaching requires careful planning for ways you can continue to support students in learning the content deeply. Taking the time to plan your lessons is an important step toward ensuring that students learn what you intend for them to learn. The outcome of a good lesson is a structured conversation among a teacher and their students about the mathematics content described in the learning goal. The plan supports teachers in tailoring the conversation to the students they are teaching in ways that focus on the mathematics they have already learned and how they learn best together.

HOW DO I PLAN MY LESSON?

One way to plan a lesson is to fill out a lesson planning template. The template below has elements that are common to many lesson plans. While filling out a lesson planning template is not required to teach well, thinking through the components of a template can support you in thinking about how best you can teach your students.

Date:	Class Period/Subject:
Big Idea(s):	Essential question and content standard
Target concepts & practices (Learning goal):	
Linking to Prior Knowledge: How does this connect to what students have already learned?	
Sharing the learning goal: What do you hope to accomplish during this lesson? (Tell the students)	
Teacher **input:** What task/problems will you assign? What notes will you give? What materials will you provide?	
Student Practice: How will students practice based on your **input?** How will you give them feedback? How will you know what individual students understand?	
Closure: How will you wrap up the lesson?	

WHAT ARE THE ELEMENTS OF A LESSON PLANNING TEMPLATE?

There are seven parts to the lesson planning template. We briefly address each of them below. The sample template has focus questions for each section to support completion.

1 *Big ideas and essential questions*: The first step of planning a lesson is reviewing the big ideas and essential questions for the unit plan you already created. Reminding yourself of the big ideas and essential questions during the planning process keeps your plan focused and ensures that the decisions you make are in service of your content goals (for more on big ideas and essential questions, see Unit Planning, p. 48).

2   *Target concepts and practices*: Next, write your learning and process goals for the lesson. These statements describe, for the teacher, what it looks like for students to learn the content of the lesson. For example, a learning goal about operating with imaginary numbers might be, "Students will use basic operations (+, −, ×, ÷) to solve equations containing imaginary numbers. They will simplify equations using substitution ($i^2 = -1$) and justify their operations using algebraic proofs (statements and reasons)."

3   *Linking to prior knowledge*: Each lesson should include an activity that connects the current lesson to mathematics that students have already learned.

4   *Sharing the learning goal*: This section is where teachers record the target concepts and practices in student-friendly language. Students are better able to assess their own understanding if they understand what the aim of the lesson is. The learning goal shared in step 2 might look this way if shared with students: "Today you will expand your application of simple operations to solve equations with imaginary numbers (i). You will use your knowledge of algebraic proofs to demonstrate your reasoning."

5   *Teacher input*: This section is where you record the task, task solutions, notes, and the materials students will use (e.g., algebra tiles). It is important that after drafting this section, you revisit sections 1 and 2 to check for alignment. It's also important that you choose a worthwhile task where students will have an opportunity to develop their own solution strategies.

Teaching in Flexible Settings

There may be times during the lesson when students need to record notes. You may choose to share an outline of your notes electronically so that students can focus on engaging more in discussions during class or provide a completed copy of your notes to students after class so they can be sure they captured the most important aspects of the discussion. See Notes, p. 139 for more on the subject.

6   *Student practice*: This section is where you record your plan for how students will practice the concepts and skills that you introduce. Record what you will be looking for in student work to indicate understanding or confusion and questions you could ask to check their understanding. A detailed learning goal is useful in developing the questions in this section (see Preparing for Discourse, p. 105 for more on how to engage in a math conversation to support student understanding).

7   *Closure*: Plan for how you will remind students of the learning goal that you shared during step 2. During the closure, students should be able to respond to the essential question for the lesson. Questions that you might ask are as follows:

- How would you describe the way you met today's learning goal?
- What does this tell us about/how does this expand our understanding of the big ideas of mathematics, such as equivalence, reflexiveness, and/or modeling?
- What opportunities did you see to make use of repeated reasoning?

Once you are done planning your lesson, revisit the story of the unit and make sure you still see the connections between this lesson and others in the unit. This review will support a cohesive learning experience for your students.

> Early in my teaching career much of my time was spent writing my lesson plans. Now I don't write out everything in physical lesson plans unless I am teaching new material. I revisit plans I've written in the past and spend time thinking through familiar topics.
>
> — HIGH SCHOOL MATH TEACHER

FREQUENTLY ASKED QUESTIONS ABOUT LESSON PLANNING

Q: Do I have to write out my lesson plan?

Find out what your district requires. Some schools don't require a lesson plan to be turned in, and some have a specific format that is required (e.g., you may need to list the statewide content standards that your lesson addresses). Talk to a peer or your mentor about what is expected of you.

Q: How do I know how long my lesson will take?

In general, you won't know how long a lesson takes until you get a feel for the students in your classroom. That said, you should aim for 10 to 15 minutes of sharing new information (teacher talk time) in a 50- to 60-minute classroom, so that students have time to work with and discuss what's been shared in the classroom with one another. If your classes are longer (80–120 minutes) you might aim for 30 minutes of teacher talk time but will want to break it up into two parts with student work time in between. For example, you may launch a task and then pause the class halfway through their work to have students share progress. You might summarize what students have found and any questions that have surfaced from more than one student in the class before having students continue their work. Remember, learning requires active engagement in the lesson, and that is easier when students don't spend an entire class period taking notes and/or listening to a lecture.

Q: How long should my lesson planning document be?

This is personal preference and based on your teaching style. If you plan to use your lesson plan while teaching, a bulleted plan might be better than a scripted one.

Q: How long should writing a lesson plan take me?

Lesson planning, like any other skill, takes time to perfect. In the beginning, aim for 30 minutes of planning time per course. If you teach more than one section of the same course, note the small adjustments you may need to make to support the students in the different courses.

Great Resource

Lesson-Planning Template from *The Mathematics Lesson-Planning Handbook, Grades 6–8.* (Williams, Kobett, & Miles, 2019). https://bit.ly/2XSwОy6

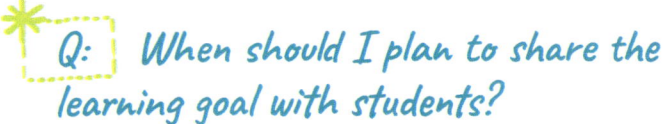

Q: When should I plan to share the learning goal with students?

Some districts require the student facing goal(s) to be listed for students to see. You may decide when to share the learning goal with your students. Sharing it at the beginning of a lesson might allow students to assess their learning during the lesson but may also unveil the discovery depending on how the task is presented. Sharing nearer to the end of the lesson might mean that students aren't able to assess their own understanding during the lesson but may allow students to assess their understanding after engaging in the lesson. Answering the essential question is another way students can assess their own learning.

Notes

What Are Mathematical Teaching Routines, and How Do I Use Them?

Instructional routines are set and repeated patterns for organizing day-to-day activities in your classroom. They can create focus and efficiency. For example, when you start class every day with an opener the students engage in while you take attendance, you establish an efficiency so that students are focused on mathematics and not on a process that isn't part of learning. Students become accustomed to the established routines, so that less time is needed on instructions, assigning groups, and other repeated parts of your lessons. For example, a routine for working in groups sets up a framework for discussions about thinking and sharing solutions and a responsibility that each student must be an active part of any group they are part of.

Routines benefit students because of their consistency—that is, students know what is expected of them—and benefit teachers because they can alleviate the need for answering innumerable questions about a classroom process, freeing the teacher's time, so they are able to focus on student work and the feedback they need to give.

MANAGEMENT

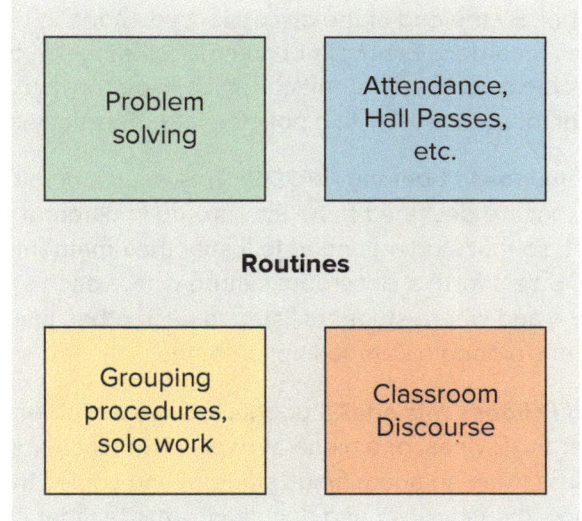

WHAT ARE EXAMPLES OF DIFFERENT ROUTINES?

Routine 1: You do—We do—I do. In this routine, students are given alone time to wrestle with a problem or task. Next, students are placed randomly into groups to share their thoughts and to think about different ways to solve the problem. Finally, after sharing out and discussion, the teacher helps the students make connections among the different approaches and generalize what has been discovered and learned. *Try—Discuss—Connect* is a similar routine in which students try a task, discuss their thinking, and then connect the different approaches to the task.

Access and Equity

These are in contrast to a traditional routine called *I do—We do—You do*, in which the teacher demonstrates a procedure or way to solve a set of problems; the students then work on a set of sample problems with the teacher and then work on their own. This does not lead to student interaction or problem solving but frequently devolves to mimicking the teacher.

What Are Mathematical Teaching Routines, and How Do I Use Them?

55

Routine 2: The four Rs—repeat, rephrase, reword, and record. This is a set of talk moves that helps students process information from classroom discussions and that helps them develop communication and language. Students are asked to *repeat* something that was said. For example, the teacher may ask a random student, "Will you repeat what Amala just said?" This is asking for a brief repetition of what was just said and reinforces good listening skills. Then students are asked to *rephrase* something from a discussion: "What did you hear the group say just now?" (which may involve combining several previous statements). The third step, *reword*, is asking students to fine-tune their language and be more precise. After the first three steps are complete, students *record* what they have learned (either in class as notes or outside of class in a journal, etc.). The recording phase can also be accomplished by the teacher scribing what the students have been saying during a class discussion of their work (*Source:* https://www.curriculumassociates.com/blog/math-instructional-routines).

Routine 3: Number talks. These are a routine students may be familiar with from their earlier learning. Students discuss an image, a mathematical expression, or something that was done quickly at their seats or in their heads. A number talk is set to be brief, maybe only 5 to 10 minutes. The teacher introduces a task and gives the students a few moments to solve/simplify/think about it. Students can signal that they are done with hand signs or some other unobtrusive method. Part of the routine is that each student has time to work and other students will not steal thinking time. The teacher then has students share solutions and records them. The teacher is a scribe as students discuss their thinking process. Students listen and ask questions, but do not correct each other. By the end of the discussion, you hope to have three or more different approaches recorded. Examples of number talks include having students solve proportion problems using different strategies (tape diagram, tables, equations) or considering a prompt such as, "Write a polynomial with exactly three real roots."

Great Resources

Christopher Danielson has a curated website with examples and resources at https://wodb.ca/

Routine 4: Which one doesn't belong (WODB). This is a set of four different answers or representations that are designed to be interpreted in different ways. Ideally, all answers are correct, so that students need to justify their mathematical thinking. The discussion for WODB sets forth a classroom culture where each student has a valid response and reason and where students listen to each other, justify their thinking, and make connections among different solution paths.

Routine 5: Why did I choose this one? Students receive a problem to solve on their individual whiteboards, devices, or a piece of paper. The teacher selects one answer to share with the class (from an anonymous student) and copies the answer for all to see. Students analyze the response and decide why the teacher chose that answer. The response may be correct and may be a typical solution, the response may be incorrect but contain a common error, or the response may be a nontypical solution path the teacher wants the students to discuss. A similar routine is *My favorite no* (https://learn.teachingchannel.com/video/class-warm-up-routine), in which the teacher selects an incorrect response so that students can find the error, but also see the value in making a mistake to aid in their learning.

Routine 6: Three reads. This helps students interpret complex word problems. *Read 1*: The teacher reads the problem aloud and asks students, "What is the big picture/What is this problem about?" (looking to understand the context). *Read 2*: Students consider, "What is the question asking? What are we supposed to figure out?" *Read 3*: Students discuss and answer, "What information do we need to answer the question? Do we need more information? How do we get started?"

What Are the Ways in Which I Can Group Students?

When students work in groups, they can discuss ideas and share their reasoning. This helps them reveal and gain a deeper understanding of the mathematics. It encourages critical thinking, justification, and argumentation skills. Learning is a social endeavor, and group work encourages engagement and reduces risk. Not all work is well suited for group work though. To get the most out of your lessons, start with a group-worthy task—one that encourages discussion and has multiple entry points, perspectives, and representations—as a foundation for student collaboration (see Tasks, p. 68). For ultimate thinking, and thus ultimate learning, you have to give students something to think about and people to think with (Liljedahl, 2021). Once you have identified what to think about (the mathematical task), how can you form groups of people to think with?

WHAT ARE THE OPTIONS FOR GROUPING MY STUDENTS?

Teachers group students in a variety of ways to make sure all students are engaged and participating in the mathematical goals they have outlined for the day. Here are some of the most common groupings:

Homogeneous grouping
Strategy: Use formative assessment data to group students together with the same or similar performances, approaches/solution strategies, or strengths.
Advantages: • Students with the same approaches, perseverance, and/or work habits can work together at their own pace. • This offers opportunities for perseverance and creative thinking for groups that process more slowly. • It offers opportunities for enrichment for groups that complete the assigned task more quickly.
Disadvantages: • The lack of diversification can stifle creativity. • Students will assign higher status to groups composed of students who work more quickly on tasks and lower status to their own groups. • High-performing students may work more independently and find the answer without sharing their reasoning with one another. This might mean that all group members are not equally certain about the solution strategy or reasoning.
Getting the most out of this grouping type: Develop norms or formative assessment strategies to ensure that all group members have shared their reasoning and understand how an answer was reached.

Great Resources

Participation Quizzes and Explanation Quizzes are good resources. See Assessments That Promote Collaborative Learning by Maika Watanabe and Laura Evans in *The Mathematics Teacher*, Vol. 109, No. 4 (pp. 298–304).

Identity and Agency

Typical roles for groups: facilitator, resource monitor, recorder/reporter, questioner from Complex Instruction; see *Strength in Numbers* by Ilana Horn for more on how to do this.

Heterogeneous grouping
Strategy: Use formative assessment data to group students together who have different performances, approaches/solution strategies, or strengths.
Advantages: • Students with different approaches, perseverance, and/or work habits can work together to benefit from the diversity by sharing different ways of thinking and doing the work. • Students with stronger content skills have the opportunity to help students with weaker content skills. • Students with stronger leadership or collaborative skills have the opportunity to facilitate the discussion to support all participants in sharing their thinking and making sense of the problem.
Disadvantages: • If the only goal of the task is to get the right answer, it is possible that the lower-performing students may take a back seat and let the higher-performing students do the work. • Students may not like who they are working with and disengage. • Students will assign higher status to individuals in the group who work more quickly on tasks and lower status to individuals who may be struggling with the material.
Getting the most out of this grouping type: Shift the purpose of the group away from answer getting and toward sensemaking of a few approaches to a problem. Develop norms around group work and/or roles that require buy-in and individual contribution.

Self-selected grouping
Strategy: Students decide for themselves what groups they want to work in based on criteria that the teacher provides, such as a given size.
Advantages: • Students can group to capitalize on the social bond and community ties in the classroom. • Students demonstrate their decision-making skills and have voice and choice to create more engagement in learning.
Disadvantages: • Students don't typically choose diverse groups and don't get a chance to work with a variety of other learners and/or personalities. • Some students are more or less sought after by their classmates. Less sought after students (or shy students) can feel isolated or embarrassed. • Students may pick groups that really should not work together due to lack of productivity or distractedness.
Getting the most out of this grouping type: Use criteria such as asking students to partner with others that push their thinking, others they have built good math working rapport with, or working with someone they have never worked with before. Develop norms that support the teacher's goals for group work.

ANOTHER TYPE OF GROUPING TO CONSIDER IS VISIBLY RANDOM GROUPING

Peter Liljedahl, in his book *Building Thinking Classrooms in Mathematics* (2021), observed from his research that visibly random grouping produced the highest amount of student engagement and thinking.

For optimum outcomes, groups should be assigned randomly, each day, and the randomness of the assignment needs to be visible to the students. Here are some aspects to consider when using visibly random groups.

- Randomized groups can be assigned in a variety of ways. Handing out playing cards to students works well because you can take one of the suits (say the hearts) and place it where that group of students will meet to work. You can call on other suits as needed: "I need the clubs to come to get the needed supplies." "I need the diamonds to meet with me for a quick clarification."
- For optimum diversity, groups of three are ideal. If a multiple of three is not available, using one or two groups of two works well because late-arriving students can join. Groups of four or more tend to be too large to keep everyone engaged.
- *Change groups every time the class meets.* Students should come to know and expect that they will be working with different classmates. At some point, all students should work with all other students in the class.
- *Differentiation looks different with randomized groups.* You will differentiate the hints and extensions you give each group based on how each group is doing. Much of the re-teaching that occurs in the classroom is done by other students within their groups.
- It may take some time for some students to become familiar with this kind of grouping, and it hinges on having really strong thinking tasks. As Liljedahl (2021) mentions in his book, you want to use "tasks that are so engaging . . . people can't resist thinking." (p. 21). Also, "they have a broad appeal, engaging contest, easy (low-floor) entry point, evolving complexity (high-ceiling), and they drive students to want to talk and to collaborate." (p. 23)

Great Resources

Noncurricular tasks resources include Math for Love; NRICH Project; NCTM problems to ponder; Julia Robinson Mathematics Festival activities; Peter Liljedahl *Building Thinking Classrooms in Mathematics* (2021); Elizabeth Cohen and Rachel Lotan *Designing groupwork: Strategies for the heterogeneous classroom*

> I started using the random number generator on my calculator to make groups several years ago. I immediately noticed that my students seemed more engaged and less likely to be off task.
>
> —HIGH SCHOOL MATH TEACHER

Great Resource

Peter Liljedahl, *Building Thinking Classrooms in Mathematics, Grades K–12: 14 Teaching Practices for Enhancing Learning.*

Teaching in Flexible Settings

Groups of five may be best when beginning work in a virtual setting. As students get comfortable with the format, groups can be made smaller.

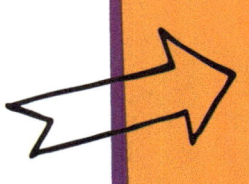

How Should My Classroom Be Organized to Maximize Student Learning?

Our classrooms should be a safe and inviting environment where student discussion and collaboration are a normal part of learning.

> Students must talk, with one another as well as in response to the teacher. . . . When students make public conjectures and reason with others about mathematics, ideas and knowledge are developed collaboratively, revealing mathematics as constructed by beings within an intellectual community.
>
> (NCTM, 1991, P. 34).

Creating this environment means considering how and where students work together and organizing your space in support of your collaboration goals.

HOW SHOULD MY CLASSROOM SEATING BE ORGANIZED?

When students walk into your classroom, they can tell a lot about how you expect them to work just by how the room appears. If desks are individually placed, students know you expect individual work. If they see two or three desks put together, they know they will work in small groups. If they walk in and see tables or pods of desks put together, they know they will work in larger groups, but not as a whole class.

Sometimes you can't help the way your classroom is designed or the furniture is set, so you may have to find workarounds. Here are some considerations in arranging your classroom seating.

Teaching in Flexible Settings

With COVID-19 safety protocols in place, classrooms may use different seating arrangements to allow for social distancing.

1 *Differ arrangements of traditional desks*: Sometimes individual seats are beneficial, and other times, you want students sitting together. If you have desks, you can keep them separated on days students are taking a test or are working on individual tasks. You can also move them together into pairs for when you want students to have a shoulder partner, or groups of threes or fours for small-group discussion.

2 *Provide multiple seating choices*: You might consider giving students the choice of different types of seats, such as yoga balls, chairs, couches, stools, tables, and so on. Not only is this flexible seating enjoyable for the students, but it promotes community and collaboration.

3 *Defront the room.* Consider how you can arrange the room so that the students view the teacher as the facilitator, not as the only source of knowledge. Liljedahl (2021) calls this *defronting* the room. You can arrange chairs, have flexible seating arrangements, and have workspaces in

different places and facing different directions, so not all students are facing the teacher, propagating the myth that the teacher is the sole source of knowledge.

Source: Liljedahl (2021).

4 *Think about who is talking to whom*: When peer-to-peer discourse is a large part of the lesson plan, U-shaped seating or circle seating allows students to face each other for whole-class discussions.

WHERE SHOULD STUDENTS WORK?

While most teachers have students sit and work at their desks, some teachers don't allow the students to have any real estate at all. They are standing and working in randomly selected groups (see Grouping, p. 57). Liljedahl's research (2021) suggests that standing produced better results in terms of student thinking than sitting because students feel anonymous when sitting, a feeling accentuated when they are sitting further away from the teacher.

When students sit at an individual desk and do their work in their own notebook, it is difficult for them to exchange ideas or questions. One way to promote cooperation and communication in the classroom is to have students work together with dry-erase markers on nonpermanent surfaces. Desktops, tabletops, windows, chalkboards, or whiteboards around the classroom are just a few surfaces that can easily be used. Dry-erase markers allow students to write and erase easily. Students start working quicker and are more willing to make and correct their mistakes if they know their work isn't written permanently (Liljedahl, 2021). As a teacher it affords you the opportunity to better see their work and progress. It also affords the students the opportunity to look around the room and learn from what others are doing.

How Should My Classroom Be Organized to Maximize Student Learning?

61

HOW SHOULD STUDENTS WORK TOGETHER?

There are a variety of ways students can work together. Depending on the goals you want to achieve, you will want to have students work together in different ways.

Pairs	Group work	Philosophical chairs	Gallery walk
Students work in pairs for some specific amount of time to process material learned, clarify directions, summarize the lesson, answer a question, give a justification, discuss ideas before sharing them, etc. A partner share approach allows more students to be participating at one time, instead of just one person responding to the teacher in typical whole-group interactions	Discussions in groups allow students to share ideas, clarify thinking, construct convincing arguments, and retain more of what they have learned Groups build a shared understanding of mathematical ideas for a deeper understanding of the material (see Grouping, p. 57, and Facilitation, p. 112)	A statement is given by the teacher that has only two possible responses: agree or disagree, true or false, yes or no Students move to the side of the room that is their response, and the two sides take turns defending their positions	Students travel from station to station to analyze other students' work or to answer questions Gallery walks can be quiet where students write feedback at each station (maybe using Post-its or writing on a posterboard) Gallery walks can also entail discussions at each station and be quite lively

What Is the Role of Practice and Homework?

Practice is an important part of math because it is a way for students to check their understanding and to learn from their mistakes. Assigning problems in class that are shared among students with self-checking gives students and teachers a sort of formative assessment of what students know and where supports are needed. Part of in-class and out-of-class work may also be assigning problems for practice of skills that have been acquired through conceptual understanding. This is a step toward achieving fluency, as students can decide which method they use to solve the problems. Problems given outside of class—that is, homework—can take several forms and purposes. Students might

1   complete problems as a check for understanding about a concept or process;
2   write up notes about what was done in class that day; or
3   answer journal prompts that allow them to explain and/or justify the steps in a problem-solving pathway, which demonstrates their conceptual understanding.

Practice doesn't have to follow directly from a skill or topic that was just taught. It may be a good idea to practice skills and concepts from earlier learning to reinforce their importance.

I've found including problems in homework that review previously learned material helps tremendously with retention. I always throw in some "just in time" review to activate the prior knowledge they will need for the next section.

—ALGEBRA TEACHER

HOW CAN I USE PRACTICE INSIDE THE CLASSROOM?

Practice can be playful, involving games and puzzles, and should have elements of choice involved. Practice may take on the role of repetition of previously learned material but should still involve active thinking and reasoning, so students get the practice they need and the motivation to sustain learning. Here are some considerations for thinking about in-class practice.

- Create a safe place for students to make mistakes as they check for understanding; consider not assigning a grade.
- The work should be done by the students, for the students' benefit (not yours).
- Practice can be done with others or independently. Overreliance on individual practice does not give value to community understanding and fosters conditions for competition versus group accomplishment.

- You are there to provide support as needed (asking questions, pressing for understanding), but be careful not to do the thinking for them (see Questioning, p. 109).
- Answers should be provided so students can self-check and revise/redo as needed.

HOW CAN I USE PRACTICE OUTSIDE THE CLASSROOM?

Homework has always been a traditional part of math class, but researchers (Liljedahl, 2021) are suggesting we need to look at homework in different ways. Homework can be a huge stress to teachers, students, and parents. Some districts require math teachers to offer homework so you'll need to be aware of the expectations of your students' caregivers, your school, and your district. Practice outside the classroom helps alleviate the time crunch experienced in class, especially short classes. It is important to think about how you will support student practice outside the classroom. Here are some points to consider when giving homework.

WHAT SHOULD I ASSIGN?

- You could give a well-chosen set of problems from the textbook or on worksheets.
- Choose websites such as IXL, Khan Academy, Desmos, Hooda Math, and Manga High; most textbook companies have adaptive software that personalizes next steps (see Technology, p. 83).
- For flipped classrooms, students will have assigned videos to watch. These students will do problems that may include practice in the classes following the videos. Videos from YouTube or Khan Academy may also be part of a homework assignment when the classroom is not in a flipped model.

HOW DO I GO OVER HOMEWORK PROBLEMS?

- Select one or two problems to go over in class. Students can vote or teachers can pick.
- Post answers to problems with solutions posted after students have had a chance to retry the problem on their own.
- Have randomly assigned groups compare their solutions to see where they agree and where they have questions.
- Avoid going over every homework problem in class. This often results in just unveiling calculation errors, which can be done in a group check or by self-checking. While a small percentage of time going over homework in some way can be useful, focusing on learning new material is usually a better use of time.

Access and Equity

Not all students have access to computers and the internet. Provide opportunities at school to get online.

> To decide what homework problems to discuss, I write all the numbers of the homework problems on the board. Students write "x" next to the problems they struggled with the most. As a class, we look at the problems with the most "x"s. If the problems have more than half the class worth of "x"s I develop a plan to re-teach those concepts.
>
> —ALGEBRA 2 TEACHER

SHOULD HOMEWORK BE GRADED?

- No. Teachers can spend an exorbitant amount of time grading. Grading homework doesn't always tell the whole story for understanding and usually tells you more about behavior than understanding. Students who consistently do homework are typically the students who have support at home or are rule followers (Liljedahl, 2021).
- When homework is graded, more worrisome behaviors appear, such as cheating off each other, getting answers from parents, or using technology (photomath, mathway, Desmos, etc.). When you are considering these behaviors, ask yourself, "Is the homework having the desired effect of helping students to acquire a more flexible and deeper understanding of a set of skills, or is it just an end unto itself to be completed?"

> I used to think giving students 30 questions a night to practice what we learned that day was good. I now realize I wasn't getting good information about what students knew. Many cheated off each other or their parents did their homework.
>
> —MIDDLE SCHOOL TEACHER

BE AWARE OF THE FOLLOWING

- Some students don't have support or resources at home to get their homework done. Assigning homework may not give you any important information about those students' understanding of the concepts and skills you are interested in. If you are assigning homework, be sure it is something that has some interest and that the time to do it is reasonable for any circumstances (multiple teachers assigning homework, responsibilities at home, etc.). Also, design homework policies that are responsive to the lives of students' cultural and family backgrounds in order to support their learning needs.
- When you assign practice problems, don't assign too many, as students who do the homework incorrectly have just then practiced the skills you want them to work on incorrectly, several times over.
- The research on the efficacy of assigning homework is mixed. While some teachers feel that homework establishes good working habits and responsibility about learning, there are no clear-cut studies that agree or disagree with this.

HOW DO I ENGAGE MY STUDENTS IN MATH?

Engagement is about how your students interact with the mathematics you want them to learn, with each other and with you. Your goals, task selection, and planning for classroom discourse are all created with the goal of having students who are engaged with their learning by solving problems, working through difficulties, sharing and explaining their thinking, justifying their conclusions, and being part of a cooperative mathematical community.

Your experience with mathematics learning most likely included many classes where you learned via the I do-we do-you do routine, in which the teacher worked sample problems while everyone furiously tried to keep up while taking notes, followed by the class doing a few problems together, and ending with each student working on their own. To have engaged students, we need to reject that routine. Be purposeful in selecting tasks that reflect your goals and that are at the proper level of cognitive demand. Position your students as active participants in their learning by helping them grow as problem solvers and critical thinkers. Let students grapple with new concepts because by engaging in productive struggle, students gain a deeper understanding of processes and ideas as well as how different ways of thinking make the concepts that undergird them clearer.

Engaging students means that each and every student is a valued part of the mathematics learning community. This means they are communicating not only with you, but with each other. Students have agency to make choices in their learning. You may feel as if you are not in total control of everything happening in the classroom. To be honest, no one ever was! Students may have appeared to be on task, but there were all sorts of off-task behavior that may have made for a quiet classroom, though not one in which all students were engaged. Rely on your routines and commonly established norms to set your classroom for engagement as well as on the tasks you have and how you account for each student.

Accounting for each student means you are using different ways to meet them where they are. Have questions and adaptations ready both for students who are struggling and for students who are completing a task relatively quickly. This ensures all your students have the opportunity to be engaged and challenged (e.g., different types of technology may allow for differentiation on the same task). Be aware of the different cultural backgrounds your students bring, and be ready to incorporate those into your planning, including addressing the needs of emergent multilingual students.

This chapter answers questions about how to engage students in the learning, which include the following:

☐ **How do I select a worthwhile task?**

☐ **How do I teach problem solving?**

☐ **How do I support my students in becoming critical thinkers?**

☐ **How do I promote and support productive struggle?**

☐ **How do I use technology?**

☐ **How do I provide differentiation for students on the continuum of prior knowledge?**

☐ **How do I support students with different learning preferences and needs?**

☐ **How do I support emergent multilinguals?**

☐ **How do I help cultivate a sense of wonder, joy, and beauty of mathematics?**

As you read about these, we encourage you to reflect on the following questions:

☐ **What does this mean to me?**

☐ **What else do I need to know about this?**

☐ **What will I do next?**

How Do I Select a Worthwhile Task?

Great Resources

For mathematical tasks, visit https://illustrativemathematics.org/, https://www.map.mathshell.org/, https://whenmathhappens.com/3-act-math/, https://www.desmos.com/, https://www.nctm.org/freeresources/

A task is a problem or collection of problems that supports students in making sense of some mathematical idea(s). A task should engage students in meaningful mathematics, have a well-defined focus, and connect to students' prior knowledge.

WHY IS TASK SELECTION IMPORTANT?

Mathematical task selection is one of the most important lesson-planning decisions teachers can make. The tasks that teachers choose influence students' ideas about what math is and what math they are capable of doing. Tasks should be engaging, motivating, and accessible for your students. While tasks should not be easy, they should be something that students feel able to attempt given their current knowledge and skills. These tasks can come from the curriculum resources that your school or district uses, and they can come from external sources as well. Both sources can be valuable.

WHAT FEATURES SHOULD I LOOK FOR IN TASKS?

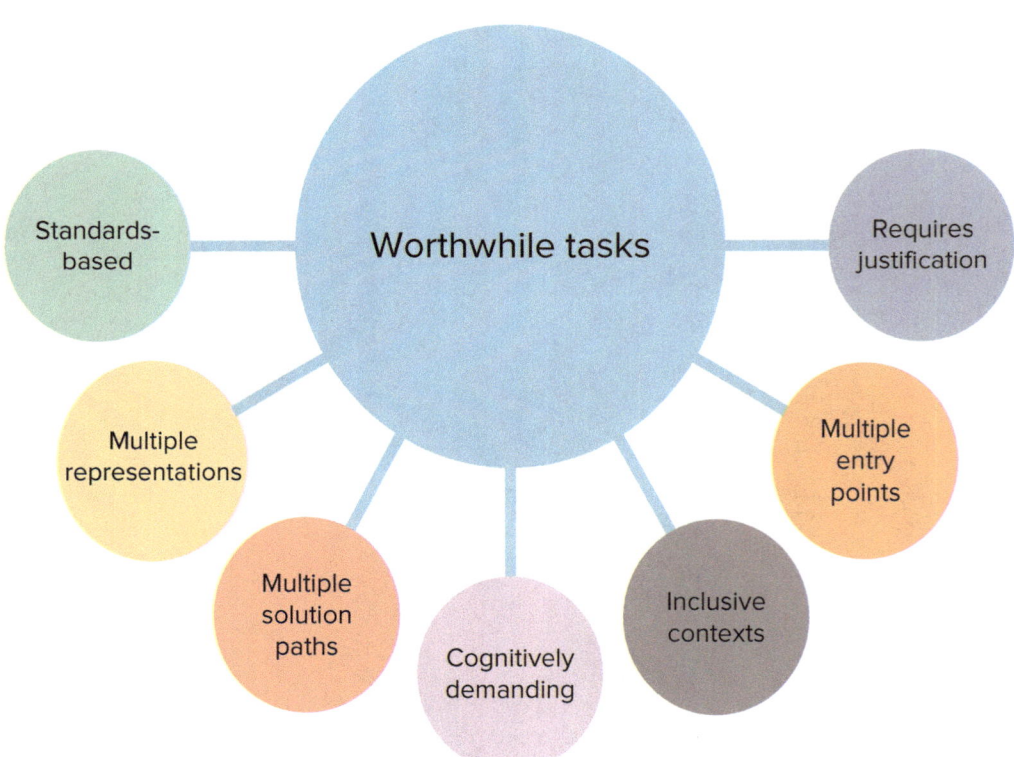

Worthwhile tasks have the following characteristics:

- *They are standards-based.* The learning goal from the lesson should guide your task selection. Learning goals are based on the content standards for the course. Teachers should choose tasks based on the concepts that they are teaching and the formative assessment data they are looking to gather (see Goals, p. 45). For example, if a learning goal for a lesson is "Students will solve equations with two variables, x and y, using substitution, grouping, factoring, and graphing," the task you select needs to provide students with opportunities to practice using those four strategies to find the solution values.

- *They use multiple representations.* Both in the presentation of the task and in the responses to the task, students should be able to see and interpret multiple representations. Lesh et al. (1987) described five types of mathematical representations that can support students' facility in making sense of mathematics. They are visual, verbal, contextual, physical, and symbolic (Principles to Actions; modified from NCTM, 2014, p. 25). Moving among these representations can support students in developing and demonstrating a deeper understanding of mathematics.

- *They have multiple solution paths.* Math problems can have one solution path and one solution (close-ended problems), many solution paths and one solution (open-middle problems), or many solution paths and many solutions (open-ended problems). Worthwhile tasks feature open-middle or open-ended problems because they provide more opportunities for students to share their thinking, engage in perseverance, and demonstrate sensemaking.

Close ended	Open middle	Open ended
Use factoring to solve for x: $(x^2 - 4) = 0$	Solve for x: $(x^2 - 4) = 0$	Develop a story problem that can be modeled by the following equation $(x^2 - 4) = 0$

- *They are cognitively demanding.* The concept of cognitive demand (Stein & Smith, 1998) suggests that mathematics is best learned when students are provided opportunities to think and reason about math concepts and make connections between those concepts and the written forms of the mathematics. Problems that do not provide these opportunities have low cognitive demand and do not contribute to conceptual understanding.

Low cognitive demand	High cognitive demand
Use the formula below to find the midpoint between (5, 6) and (11, 3): $$x_m, y_m = \left(\frac{x_1 + x_2}{2}, \frac{y_1 + y_2}{2} \right)$$	• Develop the formula for finding the midpoint of a segment by solving the problems below. Explain your reasoning: • Find the other endpoint of the segment if one endpoint is (5, 6) and the midpoint is (8, 4.5). • Find the missing coordinates. • Endpoint 1: (3, y); Endpoint 2: (x, 10); Midpoint (5, 6)

- *They use inclusive contexts.* When problems shared with students use contexts, those contexts should reflect students' identities, cultures, communities, and lived experiences as well as the vast potential of their futures (Style, 1988). This means that problem contexts should position students not only as consumers, athletes, or other subordinates but also as city planners, business owners, scientists, engineers, researchers, and so on (Rubel & McCloskey, 2021). In addition, contexts should only be used if they allow for students to engage in authentic reasoning using mathematics. If the context doesn't open space for students to reason authentically, it will diminish students' ability to make sense of the mathematics rather than enhance it (see Critical Thinking, p. 76).

- *They have multiple entry points.* Students make sense of mathematics in a variety of ways. Problems with multiple entry points support students in using many different approaches to gain access to the concepts of the problem. For example, the problem "Solve for x: $(x^2 - 4) = 0$" can be solved using factoring, graphing, or algebraically. Having these three solution paths ensures that students who don't remember how to factor a difference of squares can still approach the solution. Open-middle and open-ended problems often have multiple entry points and provide more access to the content.
- *They require justification.* The primary role of tasks in the mathematics classroom is to support students in making sense of the content. A secondary role is to support the teacher in gaining access to what students know and understand through formative and summative assessments. Tasks that require justifications make student thinking visible. The result is that the teacher has more opportunities to assess and guide student understanding.

FREQUENTLY ASKED QUESTIONS

Q: *How can I make sure that my task surfaces students' misconceptions?*

Try to include problems in your task that can uncover common misconceptions that students might have. For example, when teaching about proportional reasoning, including direct and indirect problems can help students reason about the relationship between the variables in the problem instead of blindly applying a solution strategy to find the answer.

Indirect proportion problem	Direct proportion problem
Write a story problem that accurately represents the following equation. Include any constraints/assumptions. $$7(1) = 6x$$ e.g., If it takes 1 student 7 hours to paint a mural, how long will it take 6 students? Assumption: All students must paint at the same rate.	If you own a lawn care business and must pay an arborist $300 per day, how much will you pay out if 5 arborists are working on 1 day? Explain how you found your answer using words, diagrams, or expressions.

Q: *How do I know if a task is too difficult for my students?*

If you are unable to anticipate entry points that are reflective of your students' prior knowledge or the task requires students to have memorized rather than derive formulas to find solutions, the task may cause your students significant difficulty. You should aim to give your students problems that are just beyond their capacity to stretch them, but not so far beyond that they aren't able to struggle productively (see Productive Struggle, p. 81).

Q: **How can I modify a task to make it more worthwhile?**

- Ask students to develop a story context for an expression or equation.
- Ask students to solve a problem before giving them a procedure.
- Remove some of the structures or scaffolds in the problem to allow students to apply their own.
- Have students explain their reasoning.
- Have students share their answers using more than one representation and explain how the representations communicate the same solution.
- Have students make use of mathematical structure or repeated reasoning to identify patterns in their thinking or solution strategies.
- Have students derive a formula, rule, or other generalization after working on problems (Smith et al., 2021).

Great Resource

See *On-Your-Feet Guide: Modifying Mathematical Tasks* (Eight strategies to engage students in thinking and reasoning) by Smith et al. (2021) for more on task modification.

Notes

ENGAGEMENT

How Do I Teach Problem Solving?

In the broadest terms, a problem is any situation for which an initial answer or solution strategy is not evident. If an answer is obvious, then the task is really just an exercise to practice a skill that is already known. Problem solving means more than the story problems or word problems most adults may remember from their past schooling. Problem solving should be highlighted in every topic you teach in mathematics. You know students are problem solving when they have to take care of the following:

- Decide which of their mathematical skills and tools to use on a problem (see Procedural Fluency, p. 123, and Process Standards, p. 41).
- Determine the reasonableness of their solutions (sensemaking).
- Find patterns and correlations when analyzing data (see Critical Thinking, p. 76).

The gold standard for teaching mathematical problem solving is Polya's *How to Solve It* (1945). Polya lays out a simple set of steps for addressing mathematical problems:

1. Understand the problem.
2. Devise a plan for solving the problem.
3. Implement the plan.
4. Reflect on your work.

Students may step back and reconsider as they progress.

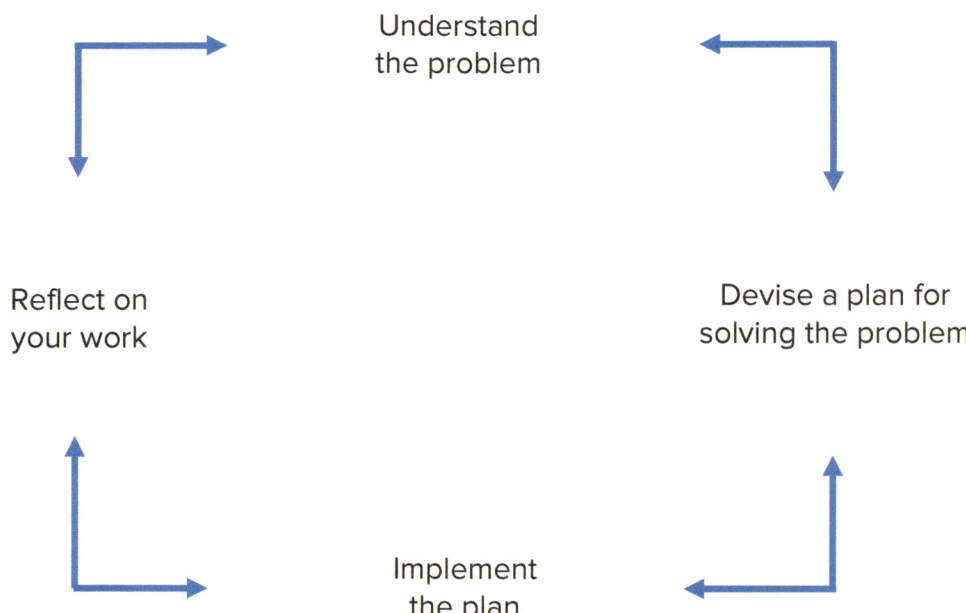

When students are asked to solve a problem, they frequently do not know how to begin. Without a game plan, they may become mired in directionless struggle, become frustrated, and then give up. With time and patience from their teachers, students develop the skills and determination to use *productive struggle* to become problem solvers. That means you provide guidance and experience solving genuine and cognitively demanding problems, so that students begin approaching problems with confidence that they can solve them (see Productive Struggle, p. 81).

The following are some tips for teaching problem solving:

- Model the behaviors you want your students to have.
- Share learning from misconceptions and plans that needed to be revised.
- Provide sufficient time for students to understand the problem and to devise a plan to solve the problem.
- Offer questions that do not take away student thinking (e.g., "What do you think would happen if the problem said Raina and Muddasir left 4 hours apart?").

WHAT ARE THE STEPS I CAN TAKE TO HELP MY STUDENTS UNDERSTAND A PROBLEM?

Begin by asking your students good questions (see Questions, p. 109). Here are some sample questions for understanding:

- What are you being asked to do or find?
- Are there any words or phrases in the problem you don't understand? How can you find out what they mean?
- What is one way you can represent the problem? (Or specifically ask for a particular representation; e.g., "How can you say this in your own words?").
- Do you need all the information in the problem? What do you still need to know?

Be careful when using strategies such as key words (e.g., "of" always means multiply) or mnemonics such as CUBES or RIDE. Giving students such a rigid plan plays into their feeling that mathematics is a set of tricks to learn. The plan becomes an end in itself, applied without looking for deeper understanding.

Consider this problem and why a strategy such as CUBES limits student understanding:

Raina drove twice the average speed as Muddasir did. Raina left 3 hours later than Muddasir. How long after Raina started had they covered the same distance?

C (circle the numbers): What should be circled as numbers? The only numeral is 3.

U (underline the question): The question includes the words how long and distance. Is this asking for a distance?

B (box the important words): What important words should be boxed? Left? Average? Later?

E (eliminate unnecessary information): What information isn't needed, if any?

S (solve and check): Does checking the solution mean considering its reasonableness?

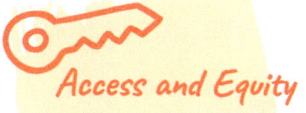

Access and Equity

Emergent multilingual students would benefit from translation into their home language and/or a display of the words and definitions in question; see Multilinguals, p. 93.

ENGAGEMENT

Great Resource

Tina Cardone's website, Nix the Tricks, gives examples of teaching without tricks but with meaning (https://nixthetricks.com/index.html).

If students have only the CUBES strategy to help them understand a problem, they will not always have a clear understanding of what is being asked. Instead, a routine such as Three Reads would be more beneficial (see Routines, p. 55).

HOW CAN I HELP MY STUDENTS DEVISE A PLAN?

Make problem solving a regular routine in your classes, so students learn different solution strategies from experience. You may want to have students keep a list of successfully used methods. Possible strategies include the following:

- Solve an easier, related problem.
- Make an ordered list of possibilities and check to see what information this gives you.
- If the problem is large or has several parts, start by solving a part of the problem.
- Write an equation that represents the problem (or make a drawing, table, or graph).
- Look for a pattern.
- Work backward.
- Think of an answer that must be wrong. How did you know that?
- Is there a formula or rule you know that you can apply?
- Is this like a problem you have solved before? How?

Great Resource

Juli K. Dixon (2020) discusses "Just-In-Time vs. Just-In-Case Scaffolding" and how that relates to equity in this article: https://bit.ly/3ENhxEn

WHAT SHOULD I DO AS MY STUDENTS CARRY OUT THEIR PLANS?

The most important thing you can do is to offer ample time for your students to understand the problem and then engage with it. Then, be patient and encourage your students to keep trying by assuring them that they have the knowledge and skills needed to solve the problem.

> "When I first started teaching, I found that I wanted to help my students. This meant I was offering too much or alerting them to possible problems before they occurred. I was stealing their opportunities to think!"
>
> —SEVENTH-GRADE MATH TEACHER, NEW YORK CITY

HOW CAN I ENCOURAGE MY STUDENTS TO REFLECT ON THEIR WORK?

Reflection time is when students think about strategies they may want to try again (adding them to their lists).

Have your students do the following:

- Write about their solving experience or discuss their pathways.

- Consider what worked, what didn't, and why.
- Look for connections among different representations and solution methods.
- Explore extensions by thinking of other ways to solve a problem after hearing the discussion.
- React to a solution you share that was not considered by the class.
- Check the reasonableness of a solution.

Also, discuss misunderstandings and what/how students learned from them. A writing prompt such as "What I Learned" can be a powerful routine to help solidify the practice of reflecting.

Great Resources

A good resource for problems is Three Act Tasks, popularized by Dan Meyer. Students start with a picture, graph, or situation that is puzzling. They notice and wonder as a way to understand the problem and decide what is being asked. Then, they think about plans for solving the problems and use their plans. Finally, they share and compare their solutions and reflect on what they did. See https://www.sfusdmath.org/3-act-tasks.html for examples.

Notes

ENGAGEMENT

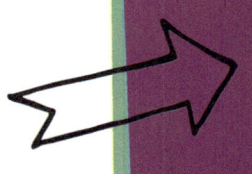

How Do I Support My Students in Becoming Critical Thinkers?

Many teachers rightly focus on helping students make sense of mathematics as sensemaking is one of the primary goals of mathematics learning. It is a necessary step to finding and checking solutions, choosing what tools or prior knowledge relates to a problem, sharing reasoning and justification of a solution, using the context of a problem to determine reasonableness of a solution, and understanding why mathematics works. While sensemaking is an important aspect of teaching mathematics, thinking critically is just as important but is not the same thing.

- Sensemaking can be defined as understanding the conceptual parts of a task and being able to share your reasoning with others coherently.
- Thinking critically is *using sensemaking* to recognize the ways math is used to understand the world outside of the classroom. For example, students can analyze patterns in data on wealth disparities and cost of living in different cities in the country to make informed decisions about what a livable wage is.

The idea of helping students make sense of the outside world may feel overwhelming, but there are some ways that process standards (NGACBP & CCSSO, 2010) can provide direction on how to do this. The following is an example of this.

Choose tasks that use real-world situations or data to have students reason about.

Great Resources

Resources for real-world tasks:

- PBS Learning Resources by Standard: https://bit.ly/39u6Qbc

- Article: Teach About Equality With These 28 NYT Graphs: https://nyti.ms/2W6CiJv

- Gapminder Tools: https://www.gapminder.org/tools/

- Census Bureau 101 for Students: https://bit.ly/3CIHGSC

- Radical Math: www.radicalmath.org

An example might be a task like this task that asks students to use visual data to compare pay gap increases by gender and by age and then relate those pay gaps to education levels by gender. Students are taught that education is the great equalizer and may believe that seniority is rewarded equally among employees regardless of gender. Having these data will support students in questioning what they have been taught to believe about the relationship between compensation and education. Students can use quantitative reasoning to make conjectures about and justify their observations on where inequities exist between employees based on gender. This task also meets mathematical content goals for reading and interpreting graphs.

Wage Gaps and Education

The U.S. Census Bureau (2019, 2021a, 2021b) shares visualizations and infographics on topics such as income, housing, and vaccinations.

Think about how you could use the graphs below to have students consider different ways to interact with and interpret graphs. Some specific types of interactions include the following:

- *Hypothesizing:* Before looking at the graphs, what do you expect to see? How do these hypotheses connect with your lived experience or the experiences of people that you know?
- *Reading the data:* Read the graphs. Do the graphs reflect your hypotheses? Why or why not?
- *Making inferences:* How might education affect the earning power of women versus men?
- *Asking yourself:* What other questions have arisen for you after looking at these graphs?

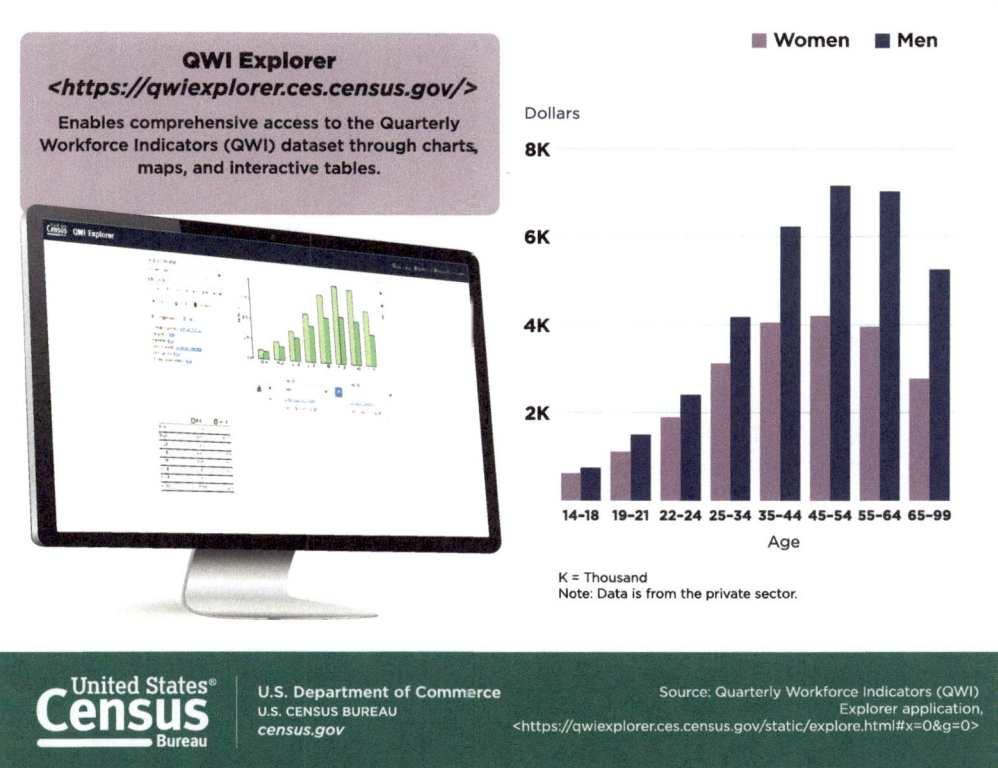

Source: U.S. Census Bureau.

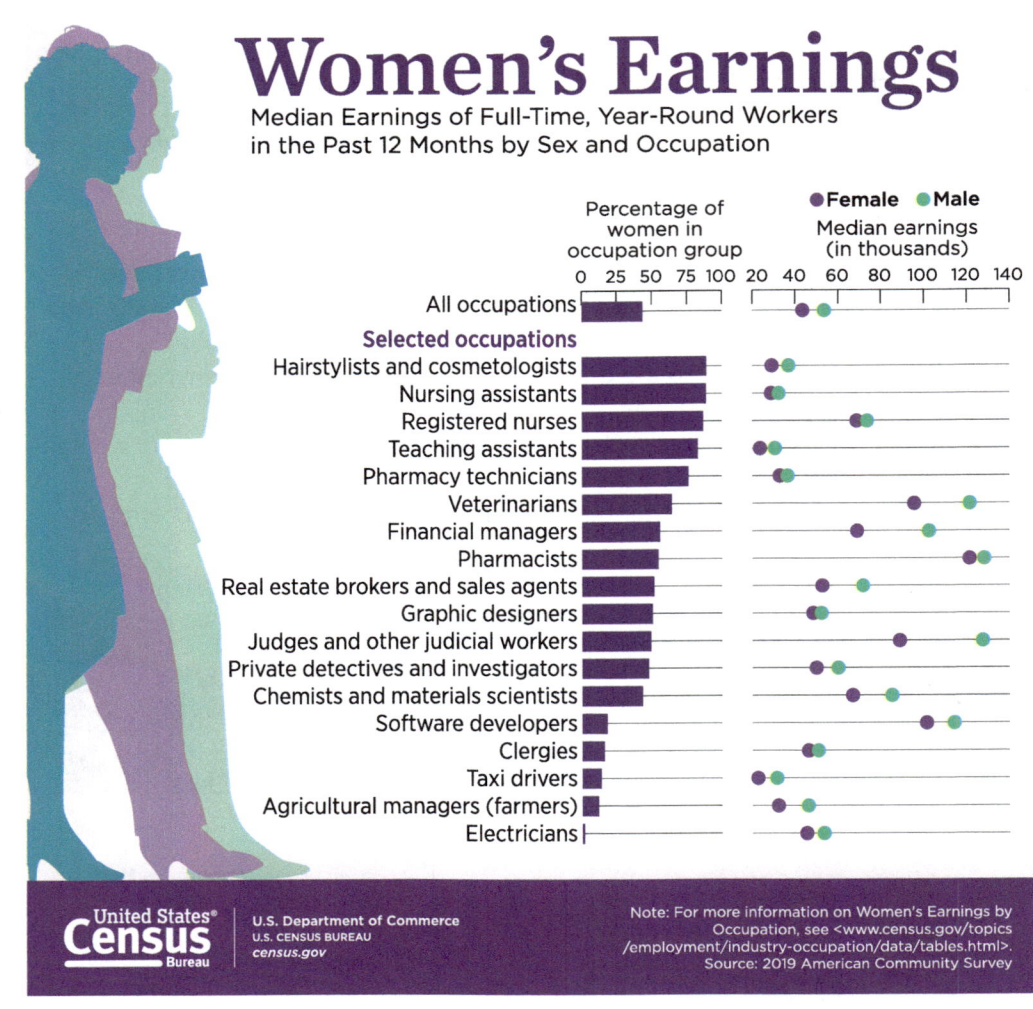

Women's Earnings

Median Earnings of Full-Time, Year-Round Workers
in the Past 12 Months by Sex and Occupation

Source: U.S. Census Bureau.

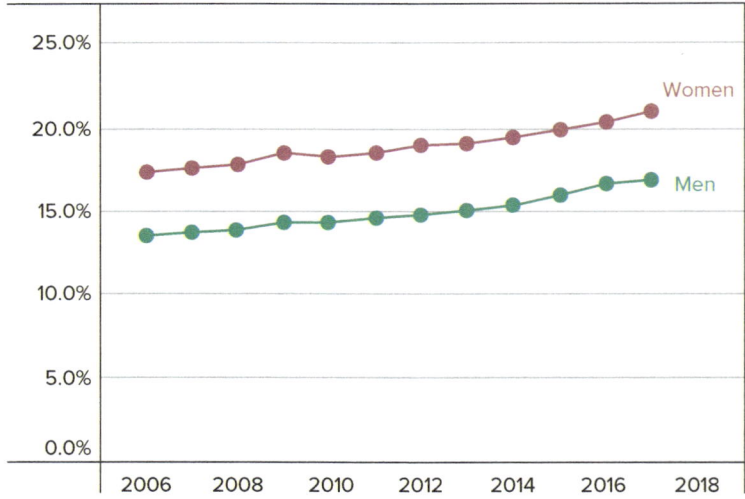

How many young adults have a bachelor's degree?

(Percentage with a bachelor's degree)

Source: U.S. Census Bureau.

Answers to Your Biggest Questions About Teaching Secondary Math

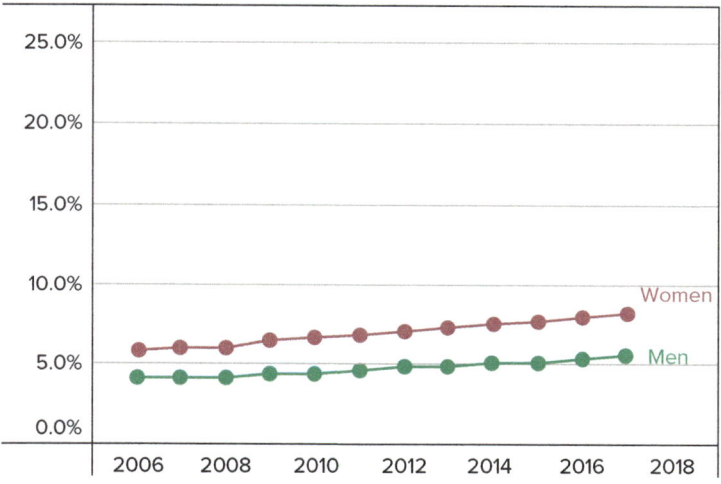

How many young adults have a graduate degree?
(Percentage with a graduate degree)

Source: U.S. Census Bureau.

Once you choose a task that highlights some aspect of students' lived experience or the lived experience of their community members, you can employ mathematical process standards to support critical thinking. For these kinds of tasks, critical thinking is best seen when students are allowed to bring their outside-of-school reasoning skills and knowledge to a mathematics task. Below are some ways to encourage that:

1 *Let students make sense of the problem:* Allow students to share conjectures about the situation based on what they know or have heard about the context. Listen carefully to how students reason about the context presented, especially when their reasoning doesn't align with your anticipated solution pathways. In the example above, this is represented by the *hypothesizing* stage.

2 *Support quantitative reasoning:* When numbers are used in a problem to make an argument, have students reason quantitatively about whether the form of quantification used could hide some truth about the solution. For example, might integers provide a view different from a proportion as a solution? Have students ask questions about the larger context and any relationships that are obscured by the way the problem is set up. In the example, the additional question prompt would be completed. Students might wonder about the data on genders not represented in the graphs or whether numbers shifted significantly after 2018.

3 *Seek to understand solutions and critique them:* Share multiple solutions with students, both those devised by other students and those that exist in the problem space, to support reasoning and sensemaking. Before students can critique other solutions, they need to make sure that they have a clear understanding of them. Share questions with students that can support them in making sense of others' solutions (e.g., How is this different or similar to my solution? Is there something that is confusing to me? How would I explain the connection between this solution and the problem space? What does this solution make me wonder?).

4 *Decide what tools or models to use:* Support students in thinking about the physical (e.g., calculator) and mathematical (e.g., proof) tools that they have at their disposal. After the conjecture phase, students should also share

what tools they think would be useful in helping them generate a solution to the problem. The tools could also include mental or physical models of the problem space that strip away some of the complexities that would affect a solution (e.g., ignoring tax when seeking livable wage).

5 *Be precise in responses:* When reasoning about real-world issues, it's important for students to know what their solutions address and what they don't. When students discuss what livable wages are, they may only be considering the impacts on employees earning a livable wage. To consider the impacts on employers or consumers, they would need to consider a different problem space. It's important that students know that each of the affected ecosystems represents their own problem space. Supporting students in this understanding can support them in thinking more critically about the conclusions they draw from their own calculations and media.

Discussing issues that are influenced by what is happening in the larger society can be daunting, but it is necessary. Consider how to provide some flexibility to students in what they study and aim to understand. One or two projects a quarter, semester, or year that allow students to communicate their curiosities and use mathematics to understand them support students' critical engagement with data, media, and popular culture. It allows for a broader treatment and application of the topics discussed in class, and it can support students in thinking in a more connected way about the mathematics that they are learning (see Culturally Relevant, p. 29).

Notes

How Do I Promote and Support Productive Struggle?

Productive struggle is an essential component of learning mathematics. Productive struggle means students work hard at a task, continue their work when faced with difficulties, and make progress. All three of these components are important. When employing productive struggle, students don't look over a task and then say, "This is too hard. I give up." Rather, they work on the task; they continue to try different solutions when the going gets tough, rather than stopping or seeking rescue; and they make progress with their efforts. The third part is a key way to differentiate productive struggle from unproductive struggle. Unproductive struggle happens when students work but do not make progress, whether because the task is not appropriate for them, or because they have no clear-cut strategies, or for any of a variety of reasons. As a teacher, you want your students to realize that they are capable of doing well in mathematics if they put forth effort and persevere.

PROMOTING PRODUCTIVE STRUGGLE

Five components combine to create the setting for productive struggle:

- 🟡 *The culture you have created in your classroom:* This must include knowing your students' mathematics identities and having a safe learning environment for each and every learner (see Community, p. 15 and Math Identity, p. 24).
- 🟡 *High expectations for all students:* This is part of the equitable structure of your class, but also means students know they are expected to try their hardest and that you and their peers are there to help when needed (and that helping does not mean telling; see Equity, p. 12).
- 🟡 *Tasks that are appropriately challenging:* This means that you need a task with a low floor, so that all students can have a starting point, but also with a high ceiling so that there is a place for everyone to be challenged (see Tasks, p. 68).
- 🟡 *Anticipating student solution paths:* Thus, you are prepared with questions and feedback that will help students when they get stuck (see Support Feedback, p. 131 and Anticipating, p. 102).
- 🟡 *Focus is on the process of finding a solution, not on getting the correct answer:* When possible, there should be multiple correct answers, but even when students may not finish the task or find a correct solution, they are still learning mathematics and perseverance (see Process Standards, p. 41).

RESPONDING TO PRODUCTIVE STRUGGLE

When faced with students who struggle, a teacher's response generally falls into one of several patterns.

Great Resource

Productive Math Struggle: A Six-Point Action Plan for Fostering Perseverance. by SanGiovanni et al., 2020.

ENGAGEMENT

Type	Characteristics	Example
Telling	The teacher corrects an error, suggests a specific strategy, gives unsolicited information.	*"You did this wrong when you added. You should have multiplied."*
Directed guidance	The teacher breaks the problem into smaller parts, asks questions that funnel students to a solution, redirects student thinking.	*"I see you are using guess and check to solve this problem. What would happen if you first made a table and then tried graphing the values from the table?"*
Probing guidance	The teacher uses the students' thinking to help them reconsider what they have done. This may take the form of asking for justification or clarification of steps, offering a hint based on student progress.	*"When you were making your diagram, how did you find the measures you used?"* (The students have an error in their diagram that their explanation may point out).
Affordance	The teacher gives little help but offers praise or encouragement.	*"I am curious to see how you continue with your use of graphs and tables. Keep trying!"*

Source: Adapted from Warschauer (2011).

SUPPORTING PRODUCTIVE STRUGGLE

Planning for and supporting productive struggle will be an ongoing part of your lessons. These are some tips that will be helpful.

- Help students realize that struggle will take time—that is, that they may not be successful, especially when they start.
- Be prepared to help students who are stuck. Ask questions such as, "What can you tell me about the problem?" or "What is an answer you know must be wrong? How do you know that?" Use questions that focus on students' thinking and the source of their confusion.
- Let students own their thinking. Do not take over the thinking for students by telling too much. Use just-in-time support (asking questions and supplying hints when needed) versus just-in-case support (trying to remove obstacles before students begin) (Dixon, 2020).
- Make sure that every student has the opportunity to struggle. Part of this will come from your task selection, while part will come from your routines for problem solving and for grouping (see Routines, p. 55, and Grouping, p. 57).
- Have available a variety of tools for students to use. This will help students think of using different representations.
- Encourage discourse and collaboration, even between groups.
- Give ample time to struggle.
- Require students to be able to justify their thinking using mathematical evidence.

Source: Adapted from Warschauer (2011).

Without productive struggle, there is no learning.

—MATH COACH

How Do I Use Technology?

Technology is all around us. It is a part of our everyday lives, and it is constantly evolving. As educators, we need to be mindful of how our students use it. We want technology to play the role of enhancing learning not replacing it. We need to think about how technology can allow students multiple entry points to interact with, explore, and understand mathematical concepts. For example, an animation for the Pythagorean theorem can show how the sum of the squares of the lengths of the sides is equal to the square of the length of the hypotenuse.

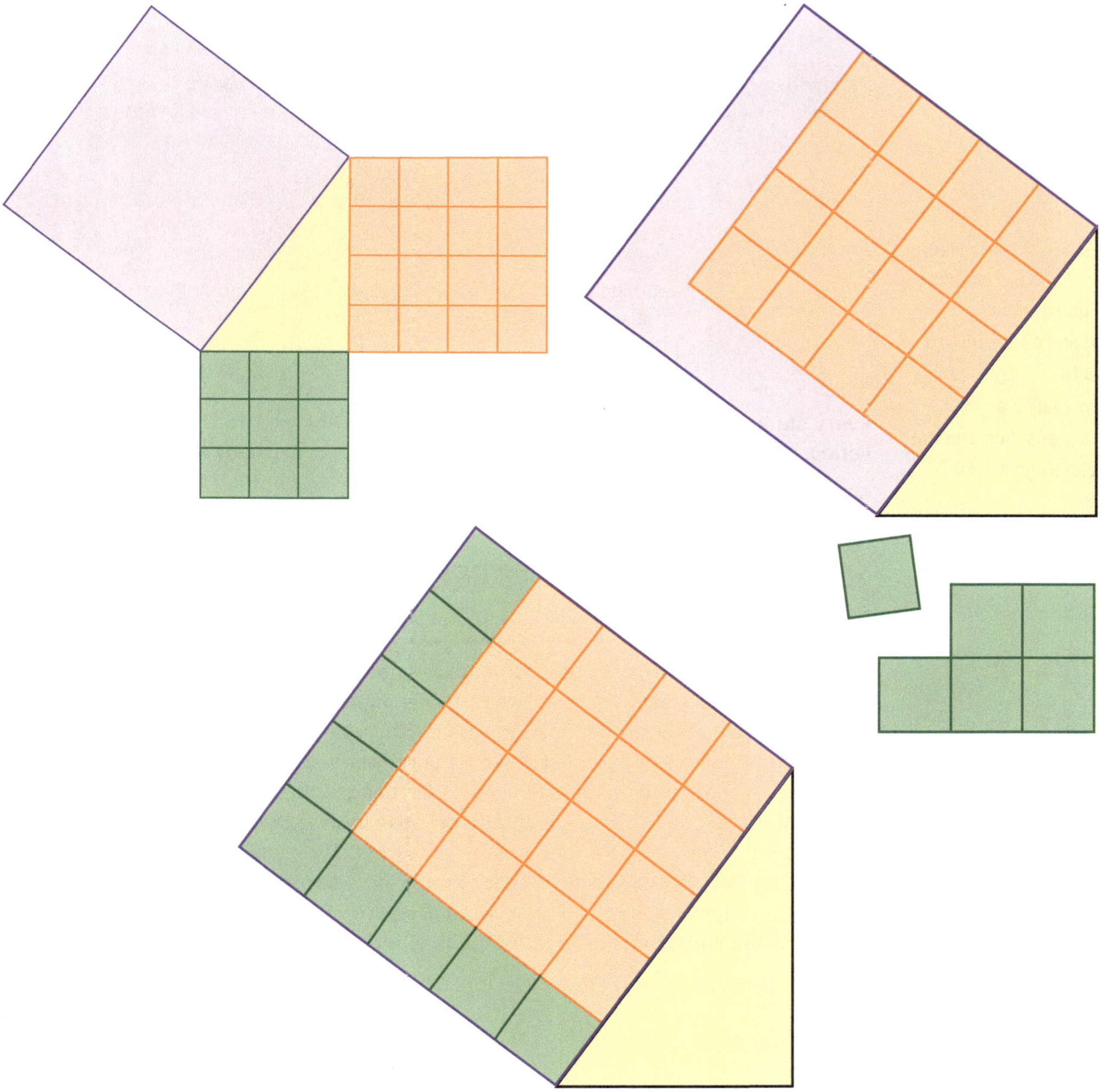

WHAT TECHNOLOGY TOOLS ARE USEFUL FOR TEACHING MATHEMATICS?

While not exhaustive, here are a variety of applications that you and your students might use to explore math.

Type	Purposes and uses	Examples
Videos and self-paced questions	Independently watch videos on various lessons and answer questions to check for understanding.	• Khan Academy • IXL • DeltaMath
Online graphing calculators	Digital tools for class activities and interactive graphing.	• Desmos • GeoGebra
Solving calculators	Math problem solvers that answer algebra questions with step-by-step explanations.	• Photomath • Mathway • Symbolab • Wolframlalpha
Interactive slides and whiteboards	An interactive presentation tool and collaborative whiteboards used to actively engage students in individual and social learning.	• Pear Deck • Explain Everything • Jamboard
Games	Game-based learning platforms that are self-paced, group-based, or whole-class oriented.	• Kahoot! • Quizlet Live • Quizzizz
Other	Learning management system, video discussion boards, and interactive video lessons.	• Google Classroom • Flipgrid • Padlet • EdPuzzle

Every student's experience with technology is different, and access for students also varies. It is important to remember that tools should be explicitly taught. Plan for how and when you will teach students to use new tools.

Teaching in Flexible Settings

You don't have the time to know every platform. What one or two tools do you know how to use and how will you use them? Consider your learning goals, unit plans, and student technology access.

HOW CAN AUTOMATIC FEEDBACK HELP MY STUDENTS?

Adaptive instructional resources for math are now evolving to generate math questions for students to practice and receive immediate feedback. Adaptive questions are questions that change with difficulty depending on student responses. For example, Khan Academy is known for supporting self-paced learners with videos, questions, and badges. Similarly, IXL changes the difficulty of the question with student responses. The advantage of these tools is that students can immediately check whether their answer is correct. They can pace themselves independently and feel gratification for practicing a specific skill. They can also figure out what skills still need more work. Desmos also allows for immediate feedback and can be customized for the teacher's needs.

HOW CAN I USE TECHNOLOGY TO DIFFERENTIATE LEARNING?

As mentioned, tools with adaptive questions can help personalize the learning experience for each student. As a teacher, you can decide what questions or skills each of your students is practicing in your class. Students also have the power and can choose what skills or questions they want to practice. Giving students time to make choices on their learning increases their engagement and retention (see Differentiation, p. 87 for more strategies).

HOW CAN TECHNOLOGY ELEVATE MY STUDENTS' VOICES?

Many of these technologies organize student responses clearly and concisely. One example is that Pear Deck or Jamboard will show all student responses as an anonymous discussion board. Teachers can highlight student responses for the entire class and begin a discussion, which allows more student voices to be heard and represented without feeling singled out. Ask your students to read another classmate's idea or even ask a question about the topic. Giving students a chance to freely share their thoughts encourages risk taking and a feeling of safety and allows students to be more open to sharing ideas. Validate as many student responses as possible when you have a discussion to encourage student ideas and thinking (see Discourse, p. 105, and Math Identity, p. 24).

HOW DO I CHOOSE THE BEST TOOL FOR MY NEEDS?

Deciding on the best tool for your students depends on you. Below is a list of questions to guide your thinking. Answer as many questions as you can.

- What is your *purpose* in using this technology?
- How do you *plan* for students to use this technology?
- What does *access* look like in your classroom for this technology?
- What are the *advantages* and *disadvantages* of this technology?
- How will you *show* students how to use the technology?
- What do you and your students *already know* about the technology?

WHAT DO I DO ABOUT PHONES AND TABLETS?

Depending on the socioeconomic makeup of your students' families, some 6 to 12 students may have personal devices such as phones or tablets. Find out your school's policy for in-class use of these devices, and think of your plan for how you will approach their use in your classroom. If you plan to have students use their devices for a lesson, be prepared to provide an option for students who may not have a phone or tablet. Touch screens can be very useful when demonstrating changes in graphs with sliders, but smaller screens can also be a limiting factor. Finally, phones and tablets are powerful small supercomputers. Photomath, Mathway, and Google Lens (even Snapchat) allow students to take pictures of math problems and solve them instantaneously. They can also allow students to take pictures of assessment questions to share with students in later classes. Be sure to consider the ramifications of the capabilities of the devices you allow.

Calculators are also powerful and helpful tools in the classroom! You get to decide how and when to incorporate them into your class. Desmos Scientific is an online calculator that is free and easy for students to use if they have access to a device. Remember, use the tools that work best for you and your students. It can be easy to get overwhelmed with the different types of technology and software available; take it one step at a time.

Access and Equity

DonorsChoose is one of several donation sites you might want to visit if there is a need for technology in your class and your school is not funded in a way to meet that need. In some states, if you require something in your class, the district must provide it. Find out your district's policy on acquiring technology for your students.

Notes

How Do I Provide Differentiation for Students on the Continuum of Prior Knowledge?

Differentiation is about providing multiple options for students to learn, make sense of their learning, and demonstrate what they have learned (Tomlinson, 2017). Teachers vary instruction and activities according to students' learning preferences, strengths, and struggles. Regardless of the method of instruction, all the students still have the same learning goal. Tailoring learning experiences to individual needs requires time to plan and commitment to implement. Differentiation could take the form of different groups answering different questions, students taking notes or getting instruction in different ways (i.e., they are using a different learning *process*), or students showing different ways to demonstrate mastery of concepts (i.e., there can be variety in the learning *product*). Refer to formative assessment to find out what your students need (see Start of the Year, p. 38, and Learning Needs, p. 90).

WHY SHOULD I DIFFERENTIATE?

Classes are filled with unique students, each of whom has their own learning preferences and experiences. In an effort to meet all their individual needs, it is tempting to assume that you must design individual lesson plans that are customized to each student's interests, strengths, and learning preferences. This is not a sustainable practice. Differentiation needs to take on a more efficient and viable process to accommodate more students' needs and varying paces of learning.

IN WHAT WAYS CAN I DIFFERENTIATE INSTRUCTION?

Tomlinson (2017) focuses on four aspects of the classroom to differentiate: content, process, product, and learning environment.

Aspect	Description	Example
Content	Adapt what students learn or how they access the new information by blending whole-class, group, and individual instruction. Students may have varying levels of familiarity with the content. Provide multiple activities for students to engage in throughout the lesson.	Solving proportions using tape diagrams, tables, graphs, and manipulatives to meet different learner needs. This also builds procedural fluency as students learn different methods for solving problems and get experience with flexibility using different methods.
Process	Students have preferred learning styles, strengths, and struggles. Get to know how your students learn best. Show students multiple strategies to solve problems or have your students teach the class their approach.	If a student came up with a way to solve a problem, name the method after that student.

(Continued)

Aspect	Description	Example
Product	Plan ahead what your students will do to show mastery for a specific learning objective. Give students flexibility in how to demonstrate mastery.	Give students the option between a traditional test or an alternate assessment/task that encompasses all the material in the chapter.
Learning environment	Adapt your classroom space for the needs of your students. Design a classroom that can allow for individual, partner, and/or group work to be changed easily.	Rearrange your classroom to have groups or stations for students (see Grouping, p. 57 and Start of the Year, p. 38).

WHAT STRUCTURES SUPPORT MY STUDENTS IN DIFFERENTIATED LEARNING?

Math in grades 6 to 12 has several tools available to support differentiated learning. The structures described below are just some of many that teachers have used to support students' learning in mathematics.

Structure 1: Technology platforms

There are several technological tools available to you to support students at varying levels of learning mathematics (see Technology, p. 83). These tools can help you customize the pace at which students go and what they learn. They can also give you more time to support students with particular needs on acquiring specific skills. For example, if you have a group of students who have mastered a particular lesson, you can use technology to allow them to learn an extended topic on their own while working with another group that is still learning the main lesson. Using this approach allows you to divide your time accordingly, so that students are challenged continuously.

Structure 2: Personalized problem sets

Create personalized problem sets for each student or groups of students. You can provide more similar problems to students who may need more practice or give students extension problems to challenge their thinking. If you assign homework, customize the assignments for students, so that they all can best achieve the learning goals for the lesson (see Homework, p. 63).

Access and Equity

When developing personalized problem sets, be sure that each set provides the student material to engage in reasoning and sensemaking. Each set should include challenging and unique problems as well. Tailoring the opportunities for learning and practice can support students. Removing opportunities for deep thinking and reasoning is an equity issue.

Structure 3: Modified assessments

Creating multiple assessments at varying levels allows you to assess students on the continuum of learning. Ensure that the assessments still assess the same learning objectives. Provide more structure or support for students who need it. For example, if you are creating an assessment that is assessing how to solve two-step equations, one version of the assessment can have single-digit coefficients, and another can have larger values (see Assessment, p. 128). In addition, some students might have specific needs according to their 504 or IEP, such as having fewer problems or number of choices (see Learning Needs, p. 90).

Structure 4: Increase the challenge

- *Extensions that dig deeper.* Visit individuals and/or groups, and give them questions that extend learning by having students dig deeper into mathematics. Sometimes just asking, "Why? Why is this happening? Does it always happen? What if you changed ___?"
- *Extensions that go farther.* Visit individuals and/or groups, and give them activities that extend learning by having students look at related but extended content. For example, now that you know how to multiply monomials, how do you think you might divide them?

Great Resources

Provide an option for students who finish the task early by posting a Problem of the Week on the wall or in the classroom. Julia Robinson Mathematics Festival (https://www.jrmf.org/), Exeter Math (https://www.exeter.edu/mathproblems), Play With Your Math (https://playwithyourmath.com/), and NCTM's Problem-of-the-Week (https://www.nctm.org/pows/) site.

- *Extensions that students create.* Have students ask their own extension questions and find their own answers—in other words, find their own extension; give different hints and extensions to different groups based on what the group has done (where they are at) and what they are ready for next.

PROVIDE STUDENTS CHOICE

1. Provide students choice in *instruction* (video, teacher lecture, reading material, group discussion with guiding questions) to learn the content.
2. Provide students choice in *showing mastery* of the standards (formal quiz or test, packet of homework or worksheet pages, use of technology to assess, project [story or poem/music/art], and verbal and/or visual presentation to teacher or class).

Great Resource

NCTM's Alternative Forms of Assessment: https://bit.ly/3u2BYlo

How Do I Support Students With Different Learning Preferences and Needs?

Students in our classrooms have several needs and learning preferences. Some students may have a particular learning disability that may make learning more difficult. Other students might need special accommodations due to a health condition. It is critical that we, as teachers, do our best to accommodate all learning needs in our classrooms.

LEARNING PREFERENCES

We all have different strengths. Students are the same! They may learn best with auditory, visual, or physical approaches. Have a discussion with your students or create a survey at the beginning of the year to learn about your students' learning preferences. After you learn more about your students, adjust your teaching practice to help as many of your students learn as possible (see Differentiation, p. 87).

INDIVIDUALIZED EDUCATION PLANS

An IEP is designed by a school psychologist and special education teacher to tailor accommodations for a student according to their documented learning disability.

Teaching in Flexible Settings

The number of students with IEPs will vary in each classroom and school. Find out how your school or district supports students with IEPs. Each school may have different approaches on how they place students into general population classes or specialized classes.

Learning disabilities are cognitive differences that a person may have that are distinct from those of the general population. Students can have a disability in one or more areas listed on the IDEA (Individuals with Disabilities Education Act) that affects their academic performance in general education. For example, a student may have a visual processing disability. This means that a student's brain may interpret what they see differently than students without this disorder. For example, they might confuse similar words, reverse letters or numbers, or have other struggles with interpreting what they see in typical ways. Your school and special education program will offer individualized services to meet a student's unique needs. As a teacher, you will be expected to adjust your curriculum, assessments, and lessons according to a student's IEP. Here are some tips to help you with those accommodations.

- **Read IEP reports before and throughout the school year**. Each IEP is updated every year and reassesses the student's disability. Reading the updated changes and accommodations will help you make better decisions about supporting your student.

- **Attend annual IEP meetings**. A psychologist, case manager (special education teacher), administrator, parent, student, and teacher meet together to discuss the student's learning progress. Case managers request teachers to come for only a part of the meeting to discuss how the student is performing in class.
- **Find student strengths and leverage those in the classroom**. Pay attention to how your student solves problems and pinpoint the strategies that work best for their learning. Building on their strengths makes students feel more confident (see Math Identity, p. 24, and Strengths, p. 21).
- **Adapt your teaching according to the student's IEP**. If a student has an auditory disability, prepare some great visuals in advance or show animations to capture how to solve problems.
- **Include and make manipulatives available**. For example, using algebra tiles to help students visualize abstract concepts can support them in their learning (see Manipulatives, p. 119).
- **Modify assessments as needed**. Some students might need to take assessments in another room or with a calculator. Other students might need an assessment with larger fonts or fewer answer choices. Prepare your assessments in advance, so that they can be adequately modified for your student (see Assessments, p. 128).
- **Communicate with caregivers regularly**. Take the time to give a learning progress update for your student's caregivers (see Communication, p. 32).
- **Ask for support**. Talk with your student's case manager or special education teacher to learn more about how to support your student. They may have learned some key strategies that are helpful in making your student more successful.
- **(When possible) Ask your student for feedback**. Talk with your student to learn about their needs. Students can often share suggestions with you to better support their learning or share strategies that have and haven't worked for them in the past.
- **Acknowledge and redirect when needed**. Some students may be challenged by maintaining focus on a task; acknowledge your students' struggles and give guided redirection to help them get back on task.

Access and Equity

If your student or caregivers speak a different language, a translator may also be present for the meeting. Some IEPs may need to be modified specifically for students who are also learning English.

ENGAGEMENT

504 PLANS

Similar to an IEP, 504 plans include accommodations to the learning environment, so that students can learn alongside their peers. A 504 plan is a blueprint for how the school will provide support for a student with a disability. In contrast to an IEP, 504 plans include *any* disability that affects the student's ability to learn in a general education classroom. Both 504 and IEP plans are provided at no cost for students. The 504 plans do not change *what* students learn but *how* they learn. These can include the following:

- Changes to the *environment*. For example, students may test in a different room or need to sit in the front of the classroom.
- Changes to *instruction*. Some plans ask teachers to check in frequently on key concepts or directions.
- Changes to how the *curriculum* is presented. For example, lessons could be provided by writing on the board or giving notes at the end of the class.

BEHAVIORAL INTERVENTION PLANS

A behavioral intervention plan is provided to students with or without an IEP or 504. It directly addresses and prevents negative behaviors, so that the student is on a positive learning track. Discuss triggers with your student and implement responsive redirection to cultivate positive behaviors in class. If students in your class ask why some students are receiving specific accommodations, speak honestly with them. Just as we would accommodate for a student who may have a broken arm in a class, we should adjust for students who may have different processing in their brains. Remind students that we respect one another despite our differences and that we accommodate, so more students succeed.

Notes

How Do I Support Emergent Multilinguals?

Emergent multilinguals are students with varying levels of command of English whose first language is not English. We refer to students as emergent multilinguals (commonly referred to as *English language learners* or *emergent bilinguals*) because they are learning more than English and often developing language fluency in multiple languages. Whether the students' families have recently emigrated or have been living in the United States for a few years, it is not uncommon for multilingual students to need language support in the mathematics class. Ultimately, any support offered to emergent multilinguals will support native English speakers as well. Let's look at some successful approaches to supporting students with varying needs in language development.

TYPES OF EMERGENT MULTILINGUALS

Emergent multilinguals in your classroom may have varying levels of time in the country. This table describes the important types of students and a short description.

Emergent multilingual type	Student's time in country	Description
Newcomer	0–2 years	Newcomers are brand new to the country. Some may have learned English in their home country, and others might not know any English at all. These students are the newest to our country's customs and schooling environment.
Developing emergent multilingual	2–4 years	Students who have been in the country for more than two years typically have enough command of English to be in mainstream classes. These students blend in with the general population since they have stronger conversational English.
Long-term emergent multilingual	5+ years	Students who speak a different language at home and are learning English at the same time as they are learning the course content. Be encouraging, patient, and supportive with respect to their academic language development.

ENGAGEMENT

Teaching in Flexible Settings

Some schools or districts offer classes/schools for newcomer students to better support them in their English and home language development. Find out what services your school or district provides for your newcomers.

WHAT ARE SOME TIPS TO SUPPORT EMERGENT MULTILINGUALS?

Supporting language acquisition in mathematics class is critical. Conveniently, mathematics is the discipline most similar across countries. The tips listed below are to help establish techniques in supporting math learning alongside English learning development. These tips are fluid and can be adapted to your emergent multilinguals' needs.

- **Partner your newcomer with a student who speaks their language**. This allows for cultural and language development to be properly supported throughout the lesson. Protect time for translation of your directions or questions for these pairs.
- **Complete an initial assessment of your newcomer's level of English and mathematics**. Set aside the time to ask your student a few questions in English to evaluate their understanding. Periodically, give your student math questions in a range of difficulty to assess their understanding and previous experience in their home country.
- When giving directions, **speak slowly, repeat frequently, and write/display the directions on the board.**
- **Read English Learning Development standards and implement them in your lesson plans**.
- **Use sentence frames when prompting questions in writing**. For example, if you ask students, "How did you solve for x?" provide the following sentence frames to help initiate student responses "First, I . . . Second, I . . . Finally, I . . ."
- **Demonstrate high expectations with your newcomers, they will surprise you!** Keep encouraging their math and language development.
- **Communicate with caregivers**. Ask your student what language their caregivers speak and arrange to find translators ahead of any conferences or conversations if needed. Your school might have translators on site or you can ask colleagues or older students to be a translator. Take the time to set this up in advance, so that you are better prepared to communicate with a student's caregiver (see Communication, p. 32).
- **Encourage translation between languages**. Students might feel more comfortable speaking in their home language. Show interest in your students' conversation (even if it is a language you don't speak), and ask students to translate for each other or for you. Translation is a high-cognitive task that should be encouraged as students continue to develop their fluency in English.
- **Be patient and help with pronunciation when students share**. Allow students to practice before sharing with the entire class; be patient and gently give suggestions for pronunciation as needed.
- **Create a safe and welcoming space**. Create a mathematical community that is welcoming to all students from different backgrounds. Make sure to greet your students at the door and group them with helpful students (see Community, p. 15).
- **Allow students to write their answers down and then practice with a partner/group *before* they share with the class**. Students learning English may be learning pronunciation or grammar and will need support from their peers (see Discourse, p. 105).
- **Display words and definitions of key academic vocabulary for a lesson or unit**. Write the words on the whiteboard or have a small poster that displays

the keywords for the unit. You could make this part of a word wall in the classroom.

- **Ask students to write *and* explain their thinking**. Support writing by asking students to write in complete sentences their answer to a question. Walk around your classroom and ask students to share what they are thinking when solving a problem. Both are critical in language development for students of all ages.

- **Validate other languages spoken and encourage translation between students**. Connect with your students by learning a few words in their home language; they will feel validated. Ask students to translate for newer students.

- **Learn more about your students' culture and home country**. Talk and listen with your students about their culture and home country; be warm to their responses and share a little about your culture too! (see Culturally Relevant, p. 29).

SHOULD I HAVE A WORD WALL?

If you teach many emergent multilinguals in your class, a word wall is a great addition to your classroom. A wall with important academic vocabulary can help with spelling and pronunciation. Students can reference it when explaining their thinking or when they need help thinking of a specific word. Asking students to write the words and draw related pictures and symbols can help the word wall support learning. In addition, you can have students write the words in their home language to support their development.

Teaching in Flexible Settings

If there is a dominant home language other than English in the classroom, be sure to learn some important phrases. Learn to count from 1 to 10, words used for basic operations, and common questions.

Notes

ENGAGEMENT

How Do I Help Cultivate a Sense of Wonder, Joy, and Beauty of Mathematics?

"Each and every student should develop deep mathematical understanding, understand and critique the world through mathematics, and experience the wonder, joy, and beauty of mathematics, which all contribute to a positive mathematical identity" (NCTM, 2020).

One of the key recommendations from the middle school edition of *Catalyzing Change* (NCTM, 2020) is to broaden the purposes of learning mathematics. Creating a meaningful mathematical learning experience involves not only access to grade-level content standards but also access to experiencing the joy and awe of mathematics. That is, math is more than just a set of benchmarks to be accomplished: math is a beautiful, wonderful, creative, and enjoyable subject. Here are some considerations in cultivating an enjoyment of mathematics:

- *See the world through a mathematical lens*. Fractals are found in the bronchial system, blood vessels, lightning bolts, coastlines, and even giant spirals of the galaxies. The Fibonacci sequence is found in the arrangement of leaves, branches, flowers, and seeds in plants. Statistical analysis plays an important role in sports. Ratio and proportion are used in art.
- *Find connections within mathematics*. Systems of equations can be solved algebraically, with matrices, or by graphing. Pascal's triangle has binomial expansion and probability applications as well as patterns such as triangular numbers, tetrahedral numbers, the Fibonacci sequence, powers of 11, and the Sierpinski Triangle.
- *Learn about history*. Learn about the history of mathematics and, in particular, the contributions of a variety of cultures to mathematics, not just Western or European contributions.
- *Value creativity*. Encourage and celebrate the creativeness of how each student solves problems.
- *Experience eureka moments*. Give time and opportunities for students to experience those magical "aha" moments when they figure something out and are proud of their accomplishments. Those moments of insight or moments of illumination create positive emotions, beliefs, and attitudes about a student's ability to do mathematics.
- *Encourage curiosity*. Have students ask their own questions and find the solutions. For example, students can create their own questions for research in statistics.
- *Use logic or problem-solving games*. For example, Nim games are mathematical games of strategy in which two players take turns removing objects from piles with the goal of either avoiding or trying to take the last object. Logic puzzles, brainteasers, riddles, and logic games engage the brain and use deductive reasoning to solve.

- *Find surprises.* Provide experiences to help students find surprising results or properties. Did you think in a crowded room two people probably share a birthday? What pattern do you notice when multiplying numbers consisting only of ones?
- *Play.* Allow students to experience the playful side of mathematics: "Mathematical play builds virtues that enable us to flourish in every area of our lives. For instance, math play builds hopefulness—when you sit with a puzzle long enough, you are exercising hope that you will eventually solve it. Math play builds community—when you share in the delight of working on a problem with another human being. And math play builds perseverance—math investigations make us more fit for the next problem, whatever that is, even if we don't solve the current problem" (Su, 2017).

LITERATURE CAN BE USED TO ENCOURAGE ENJOYMENT OF MATHEMATICS

There are a variety of books at the middle school and high school levels that invite students into the world of math.

ENGAGEMENT

Curious Incident of the Dog in the Nighttime by Mark Haddon	

Source: Doubleday books.

How Do I Help Cultivate a Sense of Wonder, Joy, and Beauty of Mathematics?

97

Flatland by Edwin Abbott	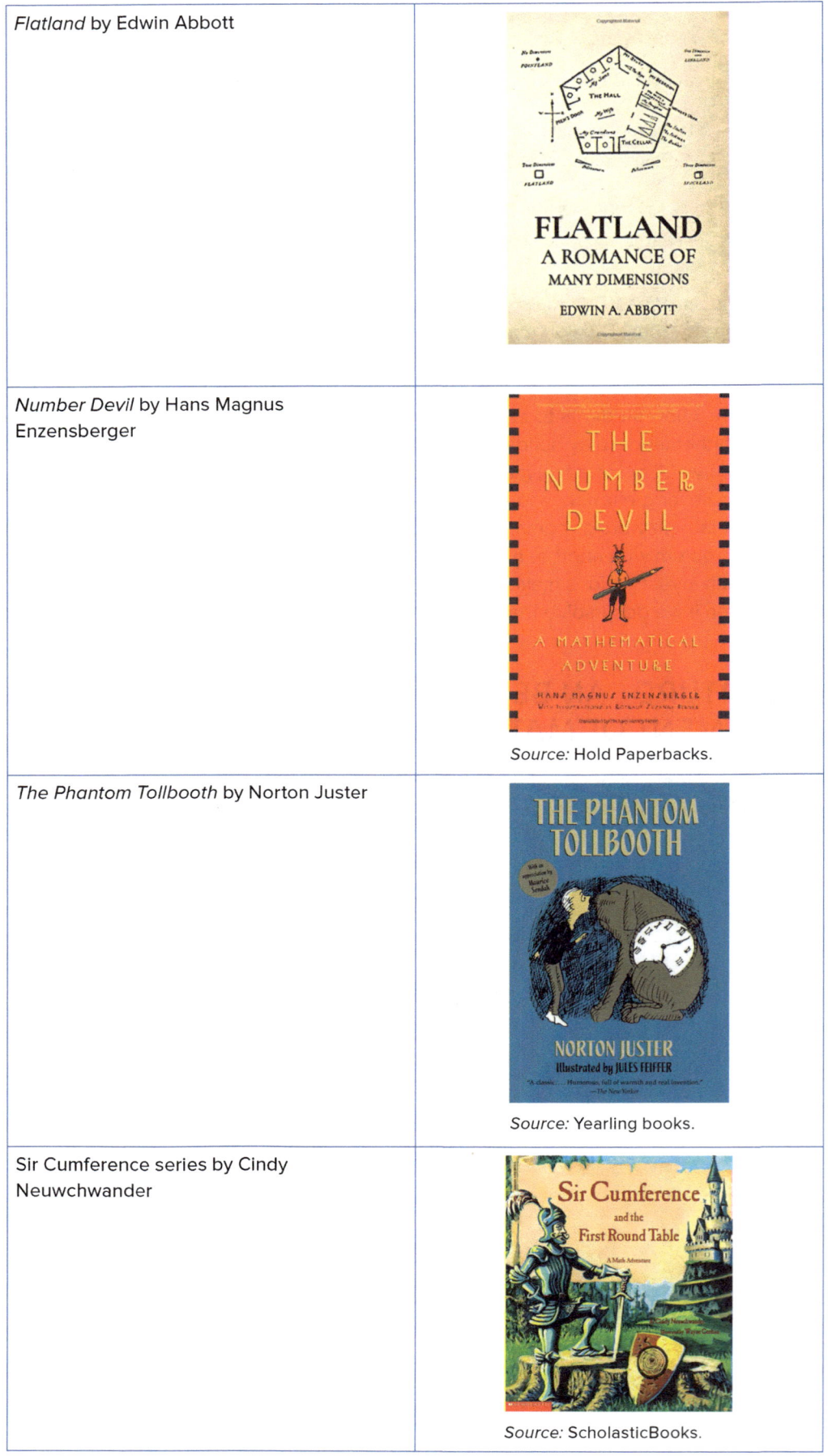
Number Devil by Hans Magnus Enzensberger	*Source:* Hold Paperbacks.
The Phantom Tollbooth by Norton Juster	*Source:* Yearling books.
Sir Cumference series by Cindy Neuwchwander	*Source:* ScholasticBooks.

The Calculus of Friendship by Steven Strogatz	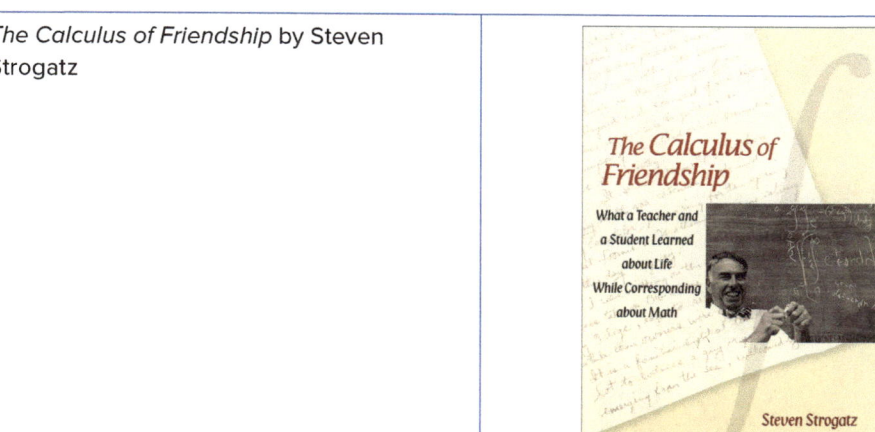 *Source:* Princeton University Press.
The Man Who Knew Infinity by Robert Kanigel	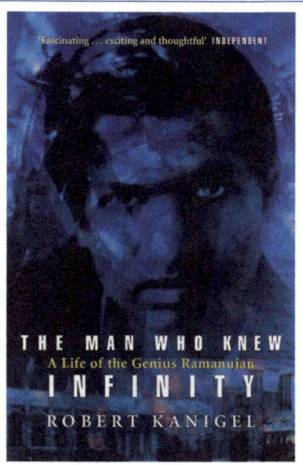 *Source:* Washington Square Press.

Great Resource

Julia Robinson Mathematics Festival at www.jrmf.org

EXTRACURRICULAR ACTIVITIES CAN BE USED TO ENCOURAGE A PASSION FOR MATHEMATICS OUTSIDE OF THE CLASSROOM

- *Start a math club.* There are many math clubs that are already organized nationally and internationally that you can join—Math Olympiad, Mathcounts, Odyssey of the Mind, and Mu Alpha Theta to name a few. Or you can create a math club to be whatever you'd like to make it—Game Theory, Mathematical Recreations, Coding Club, and so on.
- *Start a math book club.* Read a book together and discuss the book. A few books are listed above, but there is a plethora of great materials available.
- *Start a math circle.* Someone presents a problem, and the participants have fun working on it in groups and learn some new math. Check out mathcircles. org to find a circle or start your own.
- *Start a STEAM or engineering club.* Students can explore math, science, engineering, art, and math through projects, activities, and investigations.

Great Resources

Destination Imagination is an international problem-solving competition (www.imagination. org), FIRST Robotics Competition (www. firstinspires.org).

How Do I Help Cultivate a Sense of Wonder, Joy, and Beauty of Mathematics?

99

HOW DO I HELP MY STUDENTS TALK ABOUT MATH AND SHARE THEIR MATHEMATICAL THINKING?

Talking about math is part of doing math. Student discourse is one of the standards outlined in Principles and Standards for School Mathematics (NCTM, 2000) and one of the Mathematics Teaching Practices in Principles to Actions (NCTM, 2014). Students authoring ideas and communicating with one another about those ideas is an effective way to teach mathematics. Students should be engaged in worthwhile mathematical tasks that are motivating and challenging. They should freely interject ideas and strategies, describe their thinking in detail, listen to understand, and ask clarifying questions. Students should be interacting with one another and learning from each other instead of relying on the teacher.

In a student-centered classroom, teachers take on the role of facilitator instead of the sole giver of knowledge. The teacher is there to plan the lesson, make instructional decisions about implementation, and summarize learning to ensure progress toward mathematical goals. They aren't simply there to show students how to do everything and then monitor them doing it. Effective teachers believe that students are capable of being the authors of ideas and doers of the math. They support students as needed but don't take away their thinking by doing the math for them. This chapter will look at specific teacher moves to get your students talking about math and sharing their mathematical thinking.

In the book *5 Practices for Orchestrating Productive Mathematics Discussions*, Smith and Stein (2011) recommend five steps for implementing a task in a way that ensures students really get a chance to reason and develop powerful mathematical ideas through mathematical discussions. Before beginning the steps, explicit learning goals should be set and a task selected that is aligned to those learning goals. Step 1 involves *anticipating* likely student responses to the mathematical task. Step 2 involves *monitoring* students' responses and recording that information. Step 3 involves *selecting* student responses to feature during class discussion. Step 4 involves *sequencing* those responses in a way that guides students to the learning goals and makes the mathematics accessible to all students. Step 5 involves *connecting* and summarizing student responses during the discussion by using their ideas to craft the message that you want to ultimately deliver. We will dive deeper into these five practices in this chapter as well as look at the types of questions you ask and what it looks like when you are facilitating group work.

This chapter answers questions about how to help your class talk about math, including the following:

- ☐ **How do I anticipate what students will do?**
- ☐ **How do I prepare for classroom discourse?**
- ☐ **How do I plan the questions I will ask?**
- ☐ **How do I facilitate group work?**
- ☐ **How do I use multiple representations to support understanding?**
- ☐ **How do I use manipulatives in my class?**
- ☐ **What is the role of procedural fluency in my classroom?**

As you read about these, we encourage you to reflect on the following questions:

- ☐ **What does this mean to me?**
- ☐ **What else do I need to know about this?**
- ☐ **What will I do next?**

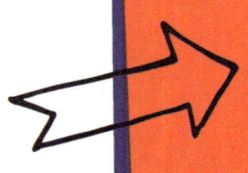

How Do I Anticipate What Students Will Do?

After you've found a worthwhile, meaningful task that meets your mathematical learning goals, start thinking about the implementation of the task. Part of your planning is anticipating how students will solve the problem and finding a way to monitor how students actually solve the problem (Smith & Stein, 2011). Anticipating involves thinking ahead about the answers and/or strategies that your students may use. Once you have anticipated, you need to decide how you will keep track of which groups used which strategies and to what degree they made progress on the task and on your learning goals. You need a monitoring tool to do this.

WHY DO I NEED TO ANTICIPATE?

We want a classroom where students are solving problems in ways that make sense to them. We also want students to develop flexibility in using a variety of methods and representations. Ultimately, we want students to make connections between and among different ways of solving problems to deepen understanding and interact with the mathematical learning goal of the day (see Connections, p. 115). Anticipating these aspects before the lesson allows the teacher the benefit of preparedness and not feeling caught off guard. It saves time in class to process what the students are doing and what your response will be. It gives you confidence in the mathematics because you've done it ahead of time.

HOW DO I ANTICIPATE?

Anticipating allows you to think about the mathematical goals you want to achieve and how those goals will be achieved through a variety of ways of solving the problem. As you think about all the different ways that a problem can be solved consider the following strategies:

- Work with other teachers and look at other resources to cast a wider net for different solution paths.
- Include correct and incorrect answer pathways as well as productive versus unproductive pathways.
- Consider not only the array of strategies but also the different ways students might interpret a problem.
- Plan questions you want to ask to assess and push each group's understanding based on the solutions paths they take.
- Think ahead of time about connections between the various strategies and how you want to make those connections explicit during discussions (see Discourse, p. 105).

Answers to Your Biggest Questions About Teaching Secondary Math

WHY DO I NEED TO MONITOR?

You have learning goals in mind and you want the class to progress in such a way that it tells a mathematical story that culminates in the learning goals being met. Thinking about students' mathematical approaches to the problem will allow you to develop that mathematical story by strategically selecting and sequencing students' work in the way you want the learning goal to unfold. Monitoring those responses will allow you to know which groups' strategies or solutions you want to share, so the story can unfold.

Monitoring is more than just walking around and looking for groups that solved the problem in particular ways. You are looking for specific solution pathways, who used them, and how deeply they developed them. When you monitor, you are asking questions to make students' thinking visible and clarifying students' thinking as well as pushing students to deepen their understanding (see Questioning, p. 109). At times you may need to ask focusing questions or give enough of a hint to help a group discover something that is an important part of the learning goal—that is, the mathematical story you want told that day.

WHAT SHOULD I RECORD AND WHY?

There are many ways to keep track of what students are doing as they explore tasks ranging from a blank clipboard where you write notes to yourself to a monitoring chart that includes not only the anticipated responses but the assessing and advancing questions you would ask each group with that anticipated response (see Questioning, p. 109). Because you will have so many things to think about during any given class period, a monitoring tool with anticipated solutions and questions to ask filled out ahead of time takes away much of the stress of "on the spot thinking." You can also sketch out a plan to sequence the anticipated solutions for the mathematical discussion to lead students to the learning goal(s) because you have thought ahead to how you want your mathematical story to progress.

Here is a sample monitoring chart:

Anticipated strategies for task:	Unit rate (per lemon):	Unit rate (per dollar):	Common price (ratio tables):		Common number of lemons:	Reasoning/ difference:	Other strategies:
8 lemons for $3, 10 lemons for $4; better buy?	8 lemons for $3 is $0.375 per lemon; 10 lemons for $4 is $0.40 per lemon.	8 lemons for $3 is 2.6666 lemon for $1; 10 lemons for $4 is 2.5 lemons for $1.	8	$3	8 lemons for $3 is 40 lemons for $15; 10 lemons for $4 is 40 lemons for $16.	10 lemons for $4 is $1 more for 2 lemons or $0.50 each. That is more than the original 8 for $3.	
			16	$6			
			24	$9			
			32	$12			
			10	$4			
			20	$8			
			30	$12			

Questions: specific to that strategy to push and probe group thinking	Why did you find the price per lemon? What does 3/8 tell you? What about 8/3?	How did you figure out how many lemons you can get for $1? What does 8/3 tell you? What about 3/8?	Explain to me your thinking. What would extending your table tell you?	Tell me more about how you solved this problem. Does this method always work?	Help me understand what you are doing here. Convince yourself, convince a friend, convince a skeptic.	
Teacher notes: Which group? What aspect to highlight?						
Sequencing: Order to best meet your learning goal	3 (Connect to ratio table for just 1 lemon.)		1	2 (Connect to ratio table for 40 lemons.)		
Teacher moves: Redirected group 4 Extended questions for group 5 Pushed group 2 for new strategy						

This is similar to the monitoring tool shared in response to the question, *How do I support student learning with feedback and grades?* (see Support Feedback, p. 131) but is more specific for group work on tasks to prepare for selecting and sequencing for a whole-class analysis and discussion. Note that this chart includes a blank row for solution paths students come up with that you never thought about. Make sure to publicly recognize every group's effort and contribution even if you don't have time to share every strategy.

WHAT DO I DO WHEN THINGS DON'T GO AS PLANNED?

Every student producing the same anticipated response or students not coming up with all the anticipated responses will happen. Here are some solutions:

- Press students to solve the problem another way.
- Give hints during the monitoring: state if groups are close or just need a little nudge to get to another one of the solution paths you want highlighted.
- Make an extension of the task (on the spot) that may open up other possibilities.
- Save sample solutions from other classes or have a specific solution available that you have worked out ahead of time.

How Do I Prepare for Classroom Discourse?

Meaningful discourse supports students in providing clear justification of their ideas and connecting their ideas to important mathematical concepts and others' solution strategies. It requires three things to be successful:

1. The implementation of worthwhile tasks
2. Adept questioning on the part of the teacher and students
3. Protected time for students to engage in the work of reasoning and sensemaking

Given these components of meaningful discourse, you can see why it takes planning and does not just *happen* in a classroom. We've discussed *anticipating* and *monitoring*, two of the *Five Practices for Orchestrating Productive Classroom Discussions* (Smith & Stein, 2011). Here, we focus on the final three practices: *selecting*, *sequencing*, and *connecting*.

Once you have anticipated what students will do with the worthwhile task you selected (see Anticipating, p. 102), you will want to (a) devise a plan for *selecting* the strategies and solutions you want to share with students, (b) decide on the order that you want to *sequence* those solutions, and (c) draft questions you can ask to support students in making *connections* between the representations you selected.

Teaching in Flexible Settings

Anticipating discusses how to make a monitoring document to keep track of students' solutions and strategies during teaching. In a hybrid or virtual setting, using a virtual monitoring document will help you track student responses and can serve as a form of classroom data (see Classroom Data, p. 136).

To demonstrate how selecting, sequencing, and connecting can be used in a classroom, we will share a possible scenario for anticipating, selecting, sequencing, and connecting strategies for a worthwhile task.

Suppose that the learning goals for our lesson are as follows:

Students will

- develop a model of a linear relationship,
- look for and express the regularity in the pattern, and
- explain what their solution(s) mean and how they connect to other representations.

To meet these learning goals, you select the following worthwhile task.

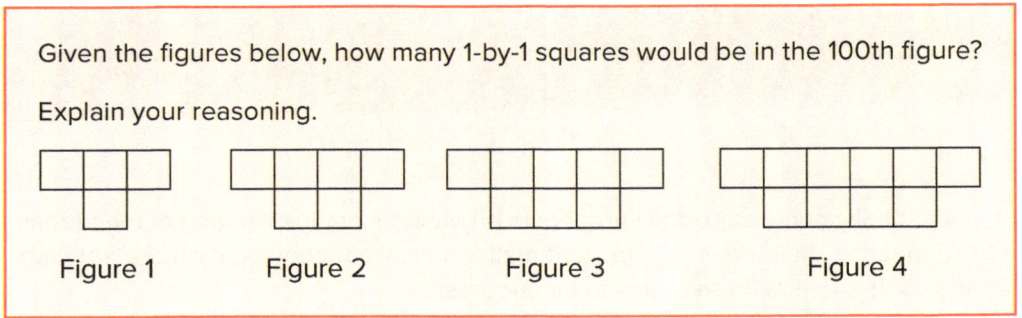

Given the figures below, how many 1-by-1 squares would be in the 100th figure?

Explain your reasoning.

Figure 1   Figure 2   Figure 3   Figure 4

Given the learning goal, students will need to see and connect multiple solution pathways in addition to finding a solution on their own.

Anticipating: You record the following solutions with justifications. While anticipating, it is important to record complete solutions as well as potential missteps, so that you notice them, either in part or in whole, when your students produce them during class. You also anticipate that students will use a solid line (as will the technology), so you prepared to discuss domain and range for the context of the problem.

Algebraic representation	**Tabular representation**		

Algebraic representation

Figure $x = 2(x) + 2$. Figure $(100) = 2(100) + 2 = 202$ blocks. Work check: Figure $(3) = 2(3) + 2 = 8$.

Visual representation

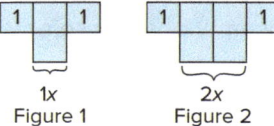

1x
Figure 1

2x
Figure 2

So, Figure 100 is 100 plus 2 more; x is 2 individual blocks. Then the 100th figure is 202 blocks.

Potential tabular misstep

Trying to multiply in the tabular representation: students might try to multiply the 10th line by 10 to get to 100. In response, have them check the relationship between Figure 2 and Figure 4.

Tabular representation

x	How x and y relate	y
1	+3	4
2	+4	6
3	+5	8
4	+6	10
5	+7	12
6	+8	14
x	$+(x + 2)$	$2x + 2$
100	+102	202

Graphical representation

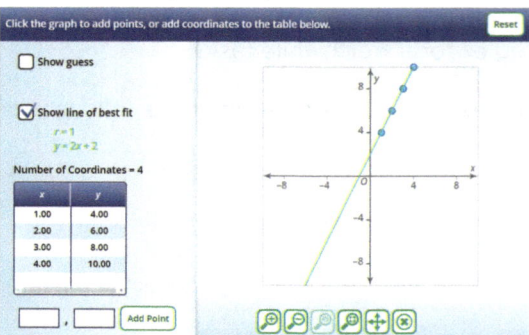

Click the graph to add points, or add coordinates to the table below.   Reset

☐ Show guess

☑ Show line of best fit
r = 1
y = 2x + 2

Number of Coordinates = 4

x	y
1.00	4.00
2.00	6.00
3.00	8.00
4.00	10.00

☐ , ☐   Add Point

Source: NCTM Resource Line of Best Fit, nctm.org.

I found the graph and the line of best, I entered 100 in as x and found y was 202. Figure 100 has 202 blocks.

HOW DO I SELECT, SEQUENCE, AND CONNECT STUDENT RESPONSES?

SELECTING

You want to select solution paths that support students in engaging in the sensemaking described in your learning goal. For this task, students need to see all four representations because the learning goal is focused on developing and connecting models of a linear relationship. Your learning goal also requires a focus on repeated reasoning, which may be a reason to highlight the tabular strategy.

SEQUENCING

There are many factors to consider when deciding how to sequence solutions and strategies. For example, you might decide to share the most visual responses first and end with the most abstract. Once you've decided on the principle for sequencing, you can decide on the order for sharing the solutions and strategies. One important consideration for open-middle tasks is that there is one answer. The name *open middle* refers to tasks that have a *closed beginning*, meaning that they all start with the same initial problem, a *closed end*, meaning that they all end with the same answer, and an *open middle*, meaning that there are multiple ways to approach and ultimately solve the problem (Kaplinsky & Johnson, 2016). For our task, if your sequencing principle is most visual to least visual, you might order the responses in this way: visual, tabular, and then graphical. This ordering has the added benefit of showing the strategies that demonstrate Math Practice 8: Look for and express regularity in repeated reasoning.

> I used to let whoever volunteered show their answer. When I read about selecting and sequencing student work, I realized how I had been cutting off the process of students thinking and sharing. Now, I plan what pathways I want students to see and in what order.
>
> —EIGHTH-GRADE TEACHER

Things to remember as you teach the lesson:

1 Be careful that you are honoring students' contributions regardless of their progression in the problem.
2 Showcasing student thinking highlights the value you have in your students' ideas and creativity. When students see that their input is valued, they feel more connected to your classroom and their classmates (see Math Identity, p. 24, and Agency, p. 27).
3 Select students who can share how they made sense of the problem.
4 Select a student who was able to restart after choosing an incorrect strategy to demonstrate perseverance and sensemaking.
5 All students will not have an opportunity to share during each lesson, but over time, all students should have a chance to share their reasoning during lessons.
6 You may choose to share only a portion of a student's strategy or solution (sharing an entire solution may be repetitive or unnecessary).

CONNECTING

Once you have settled on the solution sequence, you can begin planning how you will connect all the strategies that will be shared. Connection is built by eliciting student reasoning using questions. This is the part of the lesson where some students will be doing the work of sharing their own thinking, while all the other students are working to understand and critique that reasoning.

Here are some questions that you will want to ask:

1 What does x represent in the visual representation?
2 How is the visual representation the same or different from what x represents in the tabular form?
3 What does the graphical representation offer that you don't see on the visual or tabular forms?

Through these questions, students begin to develop the story of the lesson for themselves and progress toward the learning goal.

WHAT DO I DO IF STUDENTS AREN'T MAKING THE CONNECTIONS?

- Give students time to think, share with a partner/group, and share with the class. Record their responses publicly or build on students' thinking to build toward making connections across strategies.
- Analyze student-solved problems through error analysis. Ask students to look at pieces of work from their classmates and check for any errors. Make sure to keep the work anonymous or name it after another teacher so it does not single out a student. Students will learn to critique the mathematical thinking presented and how to correct common errors.

How Do I Plan the Questions I Will Ask?

The questions you ask your students about math, and how you ask them, play a pivotal part in your lesson. Your approach to questioning, and the questions themselves, should be driven by your mathematical learning goals and how you want students to engage in problem solving and to discuss their solution processes. There are numerous ways to classify questions and their purposes as well as a variety of routines for employing questions. Knowing these will help you plan to use them purposefully.

WHAT KIND OF QUESTIONS ARE MOST USEFUL?

There are many ways to categorize question types based on what information you want to elicit. One sorting method is a dichotomy that separates questions into two big categories of questions: *assessing* and *advancing* questions (Smith & Stein, 2011).

ASSESSING QUESTIONS

You can use *assessing questions* to understand where students are and what they know. You should plan assessing questions in advance based on the various solution pathways you anticipate students might take (see Anticipating, p. 102). Assessing questions should

- be based closely on the work the student has produced,
- clarify what the student has done and what the student understands about what they have done, and
- provide information to you about what the student understands.

> **Examples of Assessing Questions**
>
> "How did you use your graph to determine the rate of change in miles versus hours?"
>
> *Students explain different methods they may have used.*
>
> "I noticed your group was discussing this diagram. What did you notice about the lines and angles in this picture?"
>
> *Students may notice the lines intersect to make angles but may further notice whether the angles are the same size or not and whether lines are parallel, perpendicular, or neither.*
>
> "What do the parts of your equation mean?"
>
> *Students created an equation and now must explain how it fits a problem context.*

ADVANCING QUESTIONS

Once you have information from your assessing questions, use *advancing questions* to help the students deepen their understanding and/or go further in their solution pathway. You may have some general go-to advancing questions in your repertoire—such as, "Is there another way to look at this problem?" or "What would happen if *x* were changed?" But because advancing questions are based on what you're seeing from students in real time, you have to be ready to think on your feet. Advancing questions should

- use what students have produced as a basis for making progress toward the mathematical learning goal,
- move students beyond their current thinking about the mathematical ideas or strategies,
- guide students to extend what they know to a new situation to clarify mathematical ideas, and
- press students to think about mathematical ideas or strategies that they are not currently thinking about.

Source: Adapted from Smith, M. S., Bill, V. L., & Hughes, E. K. (2008).

There are other question categories that can also have a role in your lessons. Try using each type of question in your lessons, so you can learn more about your students' understanding and so students can connect to the thinking of others.

Question type	Description	Example
Gathering information	Students recall definitions, facts, and procedures.	What is a ratio? How do you graph a line if you know the slope and the *y*-intercept?
Probing thinking	Students explain their thinking, such as the steps used when completing a task.	How did you find the area of this compound shape?
Making mathematics visible	Students make connections between and among mathematical concepts, relationships, and structures.	How do a box model, a graph, and factoring demonstrate different ways to solve a quadratic equation?
Encouraging reflection and justification	Students make arguments to justify their work and think back over their work to refine, edit, and deepen their understanding.	Does your group's solution work in all situations? Convince your classmates.
Engaging with the reasoning of others	Students make connections with the work of others, and the class collaboratively comes to a deeper understanding of the mathematical ideas in a task.	How does the solution for group 2 compare with your own solution pathway? What is the same? What is different?

Source: Based on Huinker and Bill (2017).

There is a lot to consider when you are planning the questions you will use. Keep in mind the effects your questions have (e.g., justifying and engaging with the reasoning of others' questions should help foster a sense of community in your class). Good questioning techniques will grow with experience, but always require planning for your mathematical goals and the learning story you want your students to experience.

PATTERNS FOR ASKING QUESTIONS

How you ask your questions has a big impact on how students interact with mathematics and how they may come to view mathematics (e.g., assuming that there is only one way to do math versus thinking about multiple approaches). Studies have noted three prevalent routines for questioning in classes. (NCTM, 2014)

Type	Characteristics	Example
Initiate—response—evaluate (I-R-E)	• Teacher (T) asks questions with a response in mind; student (S) answers; teacher evaluates the response for correctness. • Questions are of low cognitive demand type. • Wait time is limited.	T: Who can tell me what the solution to $x + 5 = 7$ is? (5 seconds) Juan? S1: 2 T: Good job. Who can solve $3x = 6$? (5 seconds) Rae? S2: 2 T: Good work. Who can solve $10 = 8 + x$? (5 seconds) Diamond? S3: 2 T: You all are doing great work!
Funneling	• Questions set to lead to one desired solution path • Little attention to solutions that differ from the desired path • Little opportunity to connect to bigger ideas	T: How can you solve $10 = 8 + x$? S1: I drew a picture of a balance and moved blocks so the two sides balanced when the x was alone. T: Is there another way? S2: I subtracted 8 from each side. T: Good job! Why do we use that process?
Focusing	• Teacher attends to student thinking. • It is open to different approaches. • Teacher plans questions anticipating key points. • Students pressed to explain and make connections among solution paths.	T: How can we solve $10 = 8 + x$? S1: I drew a picture of a balance and moved blocks so the two sides balanced when the x was alone. T: Please share that here on the document camera. (Wait time) T: What is a different way we can share? S2: I subtracted 8 from each side. T: What do you notice about similarities and differences between the two solutions?

Using a focusing pattern should be your goal as it includes more student thinking, helps make the mathematics of solutions visible, and engages the class in comparing and contrasting different approaches, so that the mathematics task discussion is open ended and encourages divergent thinking (see Connections, p. 115).

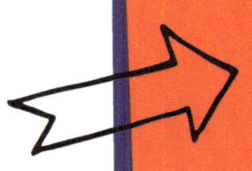

How Do I Facilitate Group Work?

As students work together in groups, encourage them to solve problems in whatever way makes sense to them, so that they are prepared to explain their approach to others in the class. As groups work and you circulate the classroom, your role as a teacher is to monitor their conversations, listen actively, and ask questions that will

- make students' thinking visible,
- help students clarify their thinking,
- ensure that members of the group are all engaged in the activity, and
- press students to consider important aspects of the task.

Your job is *not* to rescue them. While it can be difficult to maintain the cognitive demand of a task throughout a lesson, you have to resist the temptation to be overly helpful in the name of achievement or answer getting. You have to believe your students can and will think and reason at the level the task is intended to elicit. Don't take away their opportunity to think by answering questions that will shut down their thinking. Let your students answer their own questions first.

HOW CAN I ACTIVELY SEEK TO EXPLORE THE LIMITS OF STUDENTS' CURRENT UNDERSTANDING?

Start with a question such as, "Tell me more about what you have done here?" This is an open invitation to the student to explain what they did and how they were thinking. You want to ask students to explain their reasoning processes as well as their answer to gain insight into how the student made sense of the problem and assess their understanding of the mathematical ideas embedded in the problem (see Questioning, p. 109).

HOW SHOULD I ANSWER (AND NOT ANSWER) STUDENTS' QUESTIONS?

Whenever approaching a group that is asking for help, assess if they really need help or if they are just asking for help out of force of habit or learned helplessness. To figure out which questions should even be answered ask yourself the following:

- Is this a question that is being asked only because I am close by and it's convenient? Staying away from a group for the first few minutes of giving them a task will allow them to figure out what the task is asking and figure out a proposed pathway toward a solution on their own.
- Is this a question that if I answer it will cause the group to stop thinking and/or working (Liljedahl, 2021)? This type of question most commonly comes in the form of, "Is this right?"
- Is this a question that if you answer or intervene in some way the group will be able to go back to working productively (Liljedahl, 2021)? These questions, when addressed, will keep the group thinking.

If the students don't really need your help you might

- smile and walk away;
- acknowledge that they are asking a great question and encourage the group to work on figuring it out; or
- answer questions with questions—What does your group think? Does your answer make sense? Can you think of another way to solve the problem?

If the students do need your help, here are some tips to avoid rescuing or shutting down their thinking:

Lead with a question. What do you know? What have you already figured out? Many times, when kids start talking about what they do know, they end up figuring out a new path toward a solution on their own.

Explicitly connect the lesson's big ideas to what has come before and what you know will be done in the future (see Lesson Planning, p. 51).

Offer just enough information or ask just the right question so students don't have the answer but they have a way to work toward the answer. You want to provide just enough support to move beyond the sticking point while allowing for productive struggle.

Highlight important ideas for them to focus on.

Consider what you are communicating to students that you tell the answer to. You are saying, "You need me in order to succeed here. You can't do this on your own." You do not want to take over the thinking for the student by providing too much information or by "giving away" the answer or a quick route to the answer.

Make sure to follow up with the students to see what progress they have made.

WHAT SHOULD I DO IF STUDENTS ARE TAKING OVER OR NOT INVOLVED IN THE DISCUSSIONS?

Work to make sure all students have opportunities to have their voices heard. Encourage student-to-student discussions and promote positive exchanges. Asking questions such as, "Does anyone else have any other thoughts or comments?" can engage more voices to be heard. In cases where there is a student taking over the discussion, you may find it helpful to ask students to make sure two other students have had input before they are allowed to speak up again. Teachers can also roll a die to see who will answer the teacher's question about a task. This can help increase group accountability.

WHAT ARE OTHER GOOD TEACHER MOVES?

Teacher move	Rationale
Revoicing	You can either ask the student to revoice what they just said or you as a teacher can revoice. The teacher may say things such as, "So, are you saying ___" or "I'm hearing you say ___." This is a great way to summarize learning or to highlight something important that another student said.
Restating	You can ask another student to say in their own words what they just heard ___ say, or you can ask students to apply their own reasoning to someone else's reasoning. You can press students to explain why they agree or disagree with another student's work.
Wait time	Students are more willing to join in if time is provided for them to create something they feel comfortable sharing. Wait time also lets students know that you value what they think and are willing to spend the time to allow them to figure out what and how they will share.

Notes

How Do I Use Multiple Representations to Support Understanding?

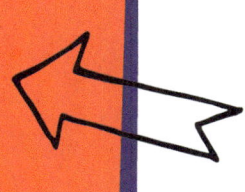

In learning mathematics, students learn more deeply when they can consider different ways to represent numbers and problems and make connections between representations (NCTM, 2014). "Representations embody critical features of mathematical constructs and actions, such as drawing diagrams and using words to show and explain the meaning of fractions, ratios, or the operation of multiplication" (NCTM, 2014, p. 24). There are three types of mathematical representations: physical representations (e.g., manipulatives), diagrammatic representations (e.g., pictures or graphs), and abstract or symbolic representations (e.g., numbers, expressions) (Dougherty et al., 2020). Lesh et al. (1987) described five common mathematical models within these types: visual, physical, symbolic, verbal, and contextual. In math courses, both tasks and solutions can be represented using one or more of these types of representations.

Oftentimes, teachers ask students to share solutions in more than one way, but teachers neglect to share problems in more than one way. Teachers may frequently assign word problems or problems using symbolic representations. While these are appropriate ways to design problems, choosing to design a problem using a graph, picture, or manipulatives can support students in approaching the problem in different ways and can clue students into the power of a representation for the purpose of sensemaking. For example, it is common for students to receive a math task using a symbolic representation such as, "Find all of the values for x for the equation $y = 20 - 2x$," but a teacher could present the problem in other ways and still reach the same mathematical goal.

Symbolic form: Find all the values of x for the equation $y = 20 - 2x$.
Contextual form: You have 20 cookies and share 2 cookies with each of your friends. Show a graph or table of this exchange and explain your reasoning.
Visual form: Develop a story problem for the following graph within the following bounds $0 \leq x \leq 10$. (The data may be discrete, but a solid line is used here for simplicity of analyzing the graph.) 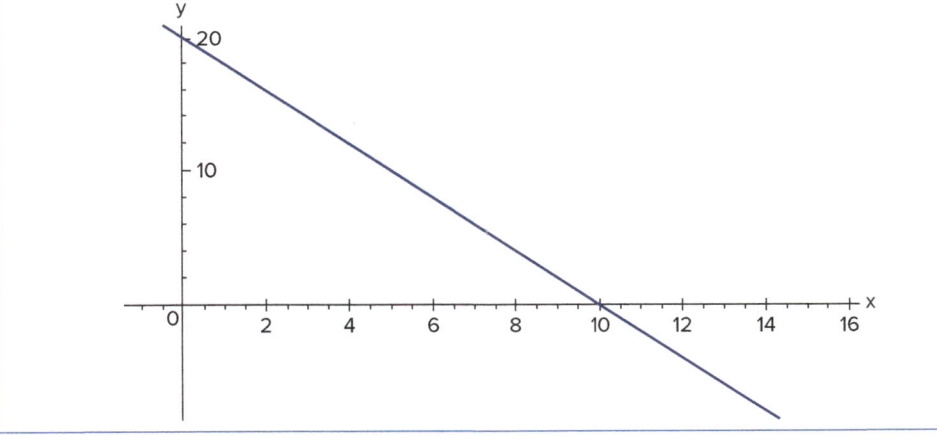

(*Continued*)

(Continued)

Visual form: Explain the relationship between friends and cookies in the table below:

Friends	Cookies
0	20
1	18
2	16
3	14
x	?

Using multiple representations allows students to approach problems from multiple perspectives, switch among representations while working toward a solution, and monitor the successes of their approaches, which creates more opportunities for successful problem solving. Students need to see multiple representations as tools to support deep and varied understanding (NCTM, 2014).

WHY IS IT IMPORTANT TO TEACH USING MULTIPLE REPRESENTATIONS?

Hammond described four ways that the brain processes new information (Hammond, 2015, pp. 132–133).

When processing new information, the brain

1. looks for similarities and differences,
2. tries to identify the larger system that the new information is a part of,
3. tries to relate the new information to prior knowledge, and
4. tries to ascertain the point of view that the new information is coming from.

While teaching and learning can provide students opportunities to process in each of the four ways, using multiple representations to teach mathematics aligns with two of the four ways of processing.

Ways the brain processes new information	Example
Similarities and differences The brain looks for similarities and differences between the new information and other familiar ideas. Using multiple representations can support students in seeing the similarities and differences between different representations of the same idea or solution.	When looking for roots of a function $f(x) = x^2 - 4$, students could show it in the following ways: 1. $x = -2$ and $x = 2$ 2. $(-2, 0)$ and $(2, 0)$ 3. Each shows the same points, but the first foregrounds the *x*-coordinate, the second and third show both x and y coordinates and make it clear that roots sit on the *x*-axis where y is zero. The location on the cartesian plane is easier to see in the graphical representations. These similarities and differences can enhance a student's understanding of what a root is and why it is a useful concept to understand.
Identifying the system or parts of the system The brain tries to organize the new information into a system. It asks whether the new information is a part of a larger system or whether it is a system that contains smaller parts. Using multiple representations supports students in recognizing systems or patterns in mathematics.	This problem uses multiple representations for a 30° angle. When learning about the unit circle, students can make sense of the patterns of the unit circle given the measures of vertical angles. Tasks like this allow students opportunities to see and use patterns to find the solutions. If $\sin(30°) = \frac{1}{2}$, $\sin(120°) = \frac{1}{2}$, $\sin(210°) = -\frac{1}{2}$, and $\sin(330°) = -\frac{1}{2}$, Find $\cos(30°) = $ ____, $\cos(120°) = $ ____, $\cos(210°) = $ ____, and $\cos(330°) = $ ____. Use your knowledge of right triangles, vertical angles, and the unit circle to support you in explaining your reasoning.
Relating new information to prior knowledge The brain tries to relate new material to familiar objects or events.	To connect factoring to prior knowledge, a teacher might ask students to do the following: Find the prime factors of 40.

(In the graph cell) X: 2.43557503   y: 1.93202575

Access and Equity

Teaching using multiple representations can support all students in accessing a task. Emergent multilinguals may have easier entry with graphical or pictorial representations, a student most comfortable with kinetic thinking may be better served by physical or virtual manipulatives, and use of an authentic context can support students in relating mathematics to skills acquired in their outside-of-school lives.

HOW CAN I SUPPORT STUDENTS' USE OF MULTIPLE REPRESENTATIONS?

Multiple representations are also useful as a tool for communication of solutions when students are solving problems. You can improve a student's ability to move fluidly between representations by trying the following tips.

Tip 1 | Assign tasks that ask for solutions to be shared in more than one way

Requesting that all students, not just those who complete tasks quickly, show more than one way to represent the solution allows students to see the connections among representations, and they can use their secondary representation to check the reasonableness of their response. One attribute of worthwhile tasks is that they have multiple solution strategies or solutions (see Selecting Tasks, p. 68, and Anticipating, p. 102).

Tip 2 | Make manipulatives available

Often, students in middle and high school are not provided with manipulatives to support them in making sense of mathematics. Ensuring that students have access to manipulatives, such as algebra tiles, patty paper, Play-Doh or molding clay, rulers, protractors, and so on, will increase the likelihood that they will use them to bolster their mathematics understanding (see Manipulatives, p. 119).

Tip 3 | Select and sequence different representations

When selecting strategies or solutions to share during whole-class discussion (see Discourse, p. 105), choose ones that represent a range of representations. Moving between representations can build a story for students about mathematical relationships. This also allows students to see the usefulness of many representations and keeps students from believing that there is only one way to solve complex mathematics problems. Over time, students will be able to identify and articulate characteristics of the representations that they work with, such as which approaches or strategies are more sophisticated, abstract, or contextualized.

Tip 4 | Support explicit comparison of the representations

As mentioned above, looking for similarities and differences is one way that the brain processes new information. Inviting students to discuss the similarities and differences among representations can train their brains to use this as a processing strategy to make sense of the mathematics. Here are some questions that you could ask students:

- What information were you able to interpret from each representation?
- What information is unique to each representation?
- What are the benefits and drawbacks of using each representation?
- How might each representation contribute to the solution process?

Using these types of questions in comparing representations with students can also support them in developing patterns of thinking during whole-class or small-group discussions about multiple representations. Modeling for students how to engage with multiple representations that are presented makes the work of sensemaking more visible.

How Do I Use Manipulatives in My Class?

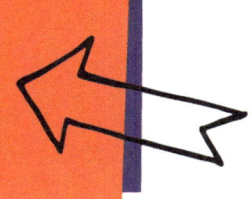

NCTM's *Principles to Actions* (2014) states that one productive belief that successful mathematics teachers hold is, "Students at all grade levels can benefit from the use of physical and virtual manipulative materials to provide visual models of a range of mathematical ideas." Manipulatives are used to help students visualize and make sense of mathematics. They are tools that help solve problems, extend understanding, transition from concrete to abstract understanding, model concepts, and show visual representations. It is easy to overlook their value if you already understand the math abstractly or are not a visual learner yourself. Middle and high school teachers tend to think manipulatives should only be used at the elementary school level, but in fact, there are a variety of manipulatives that can and should be successfully used at the secondary level.

Access and Equity

Although manipulatives are beneficial for all learners, using manipulatives is especially useful for students with learning disabilities and emergent multilingual learners.

WHAT ARE SOME APPROPRIATE SECONDARY-LEVEL MANIPULATIVES?

Here is a table with examples of how manipulatives can be used at the middle and high school levels. Check to see which manipulatives are already available in your school/district. If needed, funding is typically available through grants, or you can start a Donors Choose project (https://www.donorschoose.org/).

Type and picture	Description	Example
Pattern blocks  *Source:* Moore and Rimbey (2021).	Pattern blocks can be used for dividing fractions, area and perimeter problems, scaling, tessellations, and representing visual patterns.	How many green triangles does it take to build a scaled copy where each side is twice as long? Where each side is 3 times as long? Using a scale factor of 4?
Cuisenaire rods 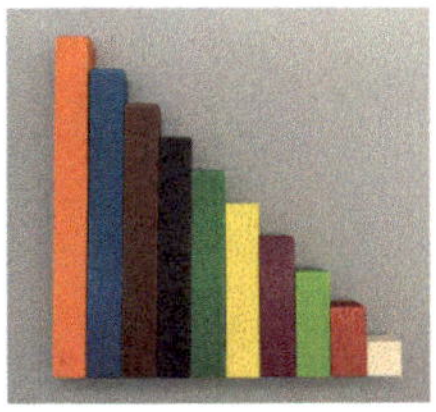 *Source:* Moore and Rimbey (2021).	Making Cuisenaire rod trains can lead to patterns with powers of two and Pascal's triangle.	How many ways can you make a train of length 3? 4? 5?
Color tiles/unit squares/grid paper 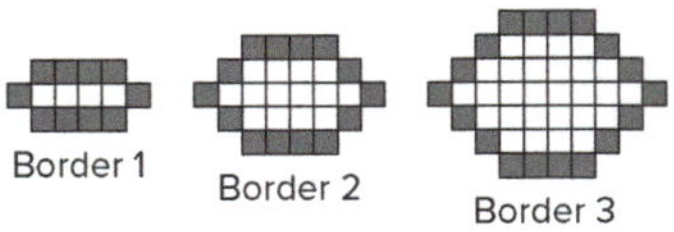	Color tiles can be used for area and perimeter problems and representing visual patterns.	How many tiles would be in border 4? 10? n?

(Continued)

Type and picture	Description	Example
Two-sided counters *Source:* Moore and Rimbey (2021).	Two-sided counters can be used to investigate integer operations.	Use counters to show how 4 + −3 = 1.
Cubes *Source:* Moore and Rimbey (2021).	Unit cubes or unifix cubes can be used for finding volume and various patterning and problem-solving problems.	If you dropped a 3 × 3 × 3 cube in some paint, how many unit cubes would have no sides painted? One face painted? Two faces painted? 3? 4? 5? 6? What about a 4 × 4 × 4 cube? 5 × 5 × 5 cube?
Dice and spinners *Die:* Nikola93/iStock.com *Spinner:* Michael Burrell/iStock.com	Dice and spinner can be used for probability and statistics experiments.	Rolling dice or spinning a spinner could simulate a randomly selected true/false answer. What are the chances of passing a 10 questions true/false test by guessing the answers?
Cutouts Radius Half the circumference	Cutouts of triangles, parallelograms, trapezoids, and circles can be used to understand formulas.	Show a visual proof of the area of a circle formula.
Algebra tiles *Source:* Moore and Rimbey (2021).	Algebra tiles can be used to understand algebraic thinking and concepts of algebra such as representing variables and constants, modeling and solving equations, multiplying, and factoring polynomials.	Multiply (x + 1)(x + 3).

Great Resource

https://solveme.
edc.org/mobiles/

Type and picture	Description	Example
Balance or mobile $5x + 1 \quad = \quad 3x + 1$ $-3x \qquad\qquad -3x$ x x • ••••• x x x x x x *Source:* MaksimYremenko/ iStock.com	A balance beam representation can be used to solve simple linear equations.	Solve $5x + 1 = 3x + 1$.
Patty paper and tracing paper	Patty paper and tracing paper can be used to learn about and experiment with transformations and properties of triangles.	Find the orthocenter of a triangle.
Unit circle on a paper plate *Source:* Image courtesy of Kristan Morales.	Something as common as a paper plate can be used to learn about the unit circle and trigonometric functions.	All sorts of common household items can be used for manipulatives: Play-Doh, marshmallows, toothpicks, yarn/string, pipe cleaners, straws, oranges, etc.
Geometric solids *Source:* Moore and Rimbey (2021).	Geometric solids can be used to investigate the relationships in polyhedral faces, edges, and vertices or volume of prisms versus pyramids. Cross-sections can also be investigated using sand or water.	Find the relationship between volume of cones and cylinders by finding how many cones of sand it takes to fill in cylinders with identical bases and heights?
Virtual manipulatives *Source:* Moore and Rimbey (2021). Created using CPM Education Algebra Tiles.	Virtual manipulatives are modeled after existing physical manipulatives. They are interactive, free of charge, and easily available online.	*Resources:* Didax Free Manipulatives Library, National Library of Virtual Manipulatives, NCTM Illuminations, Shodor Interactive Activities, Toy Theater

TIPS FOR USING MANIPULATIVES WITH YOUR STUDENTS

Tip 1 | *Use manipulatives to model a representation of a concept or skill to connect the concrete and the abstract*

- Make sure to show the work alongside. For example, if using a balance to model solving a one-variable equation, make sure to write down the way mathematicians show on paper what is being done concretely.
- Realize that this process of going between the concrete to the abstract isn't always linear, and allow students to use all representations and make connections among them.

Tip 2 | *Use manipulatives purposefully*

- To avoid confusing students, show them how to use the manipulatives with clear and simple directions.
- Some manipulatives have limitations. For example, they may only work for positive numbers or only work for integers. Be cognizant of the limitations, but realize they can still be used to find the logic of the mathematics that can then be applied beyond the limitation of the manipulative.

Tip 3 | *Make sure manipulatives are available and accessible*

- Manipulatives should be used by the student, not just the teacher "showing" a concept.
- Ensure that students know where and how to access them whenever they feel the need to use them, such as for self-directed exploration of the day's learning goal.

Tip 4 | *Allow time so students can become accustomed to using manipulatives*

- Allow time for play to acclimate to the manipulative if it is the first time a student has used it.
- Using manipulatives takes extra time, but much of that time is working toward building understanding and internalizing math processes and procedures. Time spent using manipulatives is worth it in the long run because the students take ownership and have a deeper understanding.

Tip 5 | *Set norms for caring for manipulatives*

- Set ground rules for using materials.
- Have processes in place for getting out and putting away materials.
- Remind students of the norms of the class and the responsibility they hold in treating materials respectfully.

What Is the Role of Procedural Fluency in My Classroom?

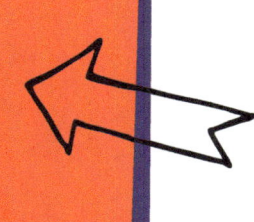

The phrase "procedural fluency" conjures images of students furiously working on long sheets of math facts in timed tests. Though many people may equate procedural fluency to being fast with calculations, it involves much more than that. Procedural fluency encompasses "knowledge of procedures, and knowledge of when and how to use them appropriately, and skill in performing them flexibly, accurately, and efficiently" (National Research Council, & Mathematics Learning Study Committee, 2001, p. 121). A key part of this quote is the words "flexibly, accurately, and efficiently." This means students have more than one process to use when approaching a problem and are able to move among those processes as needed. Additionally, knowing the steps of a procedure is meaningless unless the student can apply it accurately and with the knowledge to check their work. Efficiency can be misinterpreted to mean that a procedure is done rapidly, but that is not the case. The speed with which students apply a process is not the issue (many research studies have shown the negative effect on student identity associated with timed facts tests). Efficiency means a process is applied to a problem with understanding and without extra or wasted effort.

HOW DO I COMBINE CONCEPTUAL UNDERSTANDING WITH LEARNING PROCEDURES?

Students need to know basic facts, and they need to have procedures in place to help them be efficient mathematical thinkers and problem solvers. But procedural fluency is more than memorizing facts and processes. Several studies link an overemphasis on facts and teachers rushing students to attain fluency before they are ready to students having weaker mathematical identities because they lack a firm basis for their learning. This misunderstanding of fluency has been shown to have an adverse effect on student identity and about understanding what they have learned (Ramirez et al., 2013). For example, given a problem like $\frac{3}{x} = \frac{5}{8}$, many teachers have heard students ponder, "Am I supposed to cross multiply or cross cancel here?" signifying a lack of understanding of two procedures that were taught but not fully understood. Imagine background experiences in which these students had started studying ratios with contextual examples that were meaningful to them, and then started thinking of equivalent ratios with pictures and diagrams before using tables and the concepts of rate and unit rate. With this conceptual background for ratios, students are ready to solve proportions using the same methods and building on them by using tape diagrams, parallel number lines, and writing equations that can be solved by using inverse operations. At this point students are ready to "discover" a shortcut that we call "cross multiplying." They have a solid conceptual understanding of what ratio entails and how that relates to proportions. They have several methods to solve proportions (which will be helpful when studying related content such as percents and slope). Not only do they understand their shortcut (their efficient method), but can also explain it or re-create it if necessary.

In this case, a path that starts with meaningful contexts and uses a variety of representations to connect concrete to abstract thinking forms the foundation to understand the set of steps (i.e., the algorithm) that we know as cross multiplying. Fuson (2003) and Fuson and Beckmann (2012–2013) note that learning algorithms and fluency with understanding also decreases the occurrence of common errors and makes it less likely that students will forget an algorithm. The proportion example shows students as active learners instead of those watching the teacher demonstrate steps that they repeat rotely, increasing agency and enhancing identity (see Math Identity, p. 24).

TIPS FOR STRENGTHENING PROCEDURAL FLUENCY

Practice is important, but be mindful of how it is used. When practicing occurs before students fully understand, they may practice an algorithm incorrectly, which leads to incorrect knowledge that is difficult to correct. Practice should be brief, engaging, purposeful, and distributed (Rohrer, 2009). Keep in mind that too much practice can be ineffective or lead to math anxiety (Isaacs & Carroll, 1999).

Use writing prompts in which students explain the steps they use to solve a problem.

Have students compare different solution methods for the same problem and notice what is same and what is different, as well as which one makes the most sense to them. Similarly use a routine such as Number Talks (see Routines, p. 55) where students share their different solutions and discuss how they compare. Students can also compare solutions with common errors to identify them and think of ways to avoid them.

Include tasks that demonstrate how an algorithm is both efficient and powerful in that it can solve an entire class of problems. For example, using a box model to multiply two-digit numbers is the same as using a box model to multiply complex number expressions such as $3 + 2i$ times $4 - 5i$, and is the same as multiplying binomials. Additionally, flexibility becomes important when students think of algorithms that don't require physically drawing a box that use the same reasoning, even if the steps look different (e.g., the standard algorithm, the double distributive property, or Each One Gets One).

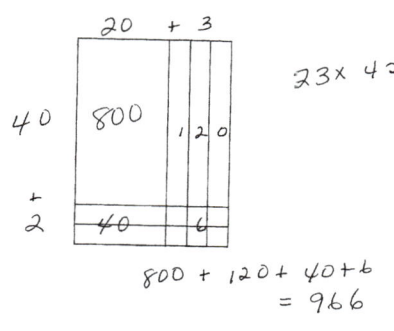

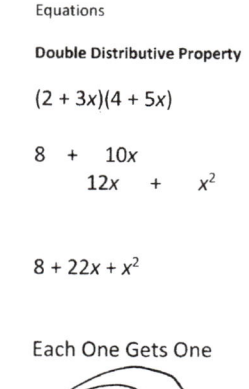

$$23 \times 42$$

Equations

Double Distributive Property

$$(2 + 3x)(4 + 5x)$$

$$8 \quad + \quad 10x$$
$$\quad\quad 12x \quad + \quad x^2$$

$$8 + 22x + x^2$$

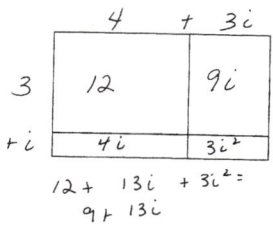

Each One Gets One

$$(2 + 3x)(7 + 4x + x^2)$$

$$14 \quad + \quad 8x \quad + \quad 2x^2$$
$$\quad 21x \quad + \quad 12x^2 \quad + \quad 3x^3$$

$$14 + 29x + 14x^2 + 3x^3$$

The connections among the different solutions are a perfect example of why conceptual understanding is important as students adapt an algorithm for one topic to other topics. Without conceptual understanding, students may see the multiplication processes as isolated bits of knowledge, and thus compartmentalize them so that they think different problems such as multiplying two binomials is very different from multiplying a binomial and a trinomial, even though the processes are very similar.

Tip 5

Provide students with opportunities to use their own reasoning strategies when solving problems. Then have them explain their strategies and help the class make connections between different strategies to determine which makes the most sense and is most efficient for them (W. G. Martin, 2009).

Great Resources

Klerlein, J., & Hervey, S. Mathematics as a complex problem-solving activity. *Generation Ready*. https://www.generationready.com/white-papers/mathematics-as-a-complex-problem-solving-activity/

Fernandez, M. L., Hadaway, N., Wilson, J. W., & Graeber, A. O. (1994, March). Connecting research to teaching: Problem solving: Managing it all. *Mathematics Teacher, 87*(3).

Centre for Teaching Excellence, University of Waterloo. *Teaching problem-solving skills*. https://uwaterloo.ca/centre-for-teaching-excellence/teaching-resources/teaching-tips/developing-assignments/cross-discipline-skills/teaching-problem-solving-skills

HOW DO I KNOW WHAT MY STUDENTS KNOW, AND HOW CAN I USE THAT INFORMATION TO PLAN AND MOVE FORWARD?

Displaying proficiency in mathematics takes many forms. Your students will showcase their mathematical thinking verbally, visually, and/or in writing. You will gather data on student thinking and give feedback to support them on their path to mastery. This information helps you learn the best ways to assess your students' growth and progress over time. Analyzing the data gives you a better understanding of your students' learning progress. Furthermore, it will improve your practice because it will be more in tune with how your students are learning. Afterward, planning is adjusted for students to deepen their mathematical understanding.

This chapter investigates assessing student knowledge and adapting planning to meet your student's needs. First, we discuss the difference between *formative and summative assessments* to determine how students are demonstrating mathematical mastery. Assessing student learning in real time, before and after a lesson, reshapes your teaching by highlighting student strengths and needs. Finally, we discuss how to give students feedback during or after a lesson. Properly constructed feedback reinforces and develops mathematical thinking. Then, dive deep into how you are *collecting and analyzing data* that you receive on a daily basis in your classroom. Several decisions are made both in the moment and during the planning stages to improve student learning progression. Finally, we discuss the use of student notes to organize their thinking and process mathematical problem solving. Notes encourage students to summarize and consolidate multiple problem strategies.

This chapter answers questions about how to know what your students know and use that information to move forward and includes the following:

- ☐ **How can I use information from formative and summative assessments?**
- ☐ **How do I support my students with feedback?**
- ☐ **How do I analyze classroom data?**
- ☐ **What is the role of notes?**

As you read about these, we encourage you to reflect on the following questions:

- ☐ **What does this mean to me?**
- ☐ **What else do I need to know about this?**
- ☐ **What will I do next?**

How Can I Use Information From Formative and Summative Assessments?

Finding out what your students know before a lesson is helpful when planning. Additionally, determining what your students know during a lesson allows you to adjust your teaching as the lesson progresses. Finally, understanding what your students know after the lesson shows you how well your students succeeded in attaining your mathematical learning goals. All these components are part of your assessments of your students. The information you learn from your assessments informs all your teaching decisions.

WHAT IS SUMMATIVE ASSESSMENT?

Assessment is generally split into two types: formative and summative. When most adults think about their learning experiences, they focus on summative assessments because those are what are used for measurements of standards (e.g., statewide testing) or for grading purposes. Summative assessments tend to be done at the conclusion of a set of learning goals, usually as a quiz or test, as a cumulative exam (e.g., a semester final). Summative assessments also include standardized tests such as the SAT or ACT, or a long-term performance assessment, such as a portfolio or a capstone project (e.g., an International Baccalaureate project). Summative assessments are largely done to evaluate learning, as opposed to being used diagnostically. In your classroom, summative assessments will probably be used to show how students have mastered your learning goals (procedural fluency on skills, conceptual understanding, problem solving, etc.) for a specified time period. Your summative assessments may be used as a basis for some diagnosis because you may change your plans for instruction based on the outcomes you see.

> ### Summative assessment tips
>
> It's okay to use existing summative assessments, maybe from your texts, as you grow your understanding of your curriculum.
>
> You should have a draft summative assessment completed before you start teaching the material, so your goals are clear to you (see Unit Planning, p. 48).

Large-scale summative assessments may give you a broad overview of your school's programs. If your district results show students are not meeting a state-level standard, that gives you an indication about where you might focus. But be aware that test scores without context can be misinterpreted. The effects of socioeconomic status and/or inherent cultural and social biases are just two contexts that must be considered when analyzing large-scale summative assessment results.

WHAT IS FORMATIVE ASSESSMENT, AND HOW CAN I USE THE INFORMATION I GAIN FROM THAT?

Formative assessments are used to take a snapshot of student learning and understanding and then to inform teacher decisions about instruction. Formative assessments happen throughout a class period and range from informal observations to specific questions. The following are some types of formative assessments that you can consider using.

Type	How to use
Observation	Watch for key things that you have anticipated and for solid understandings and different misunderstandings.
Hand signals	Use three different hand signals with your students. With their hands next to their chests, have students signal thumbs up for "I understand this," flat hand for "I have a question" or "I am unsure about this," and thumbs down for "I do not understand this." Anonymity is an important part of this, so that students are not afraid to express how they truly feel.
Hinge questions	Hinge questions are questions that are based on key concepts from the lesson. They should have answers that are quick so you can determine if students have grasped the concepts before moving on. Try to start with one or two in a lesson, usually at a midway point, as you practice using these.
Personal whiteboards	Students write their work on their own board that they hold up for you to see.
How did you do this?	Have a student or group of students explain how they did something rather than demonstrating a skill, so you have a better picture of what they understand.
Interviews	Ask questions prepared to elicit information about student understanding of a concept (see Questions, p. 109).
Technology (see p. 83)	Kahoot is a good example of a whole-class app that you can use to see how students are progressing.
Four corners	Given a topic about mathematical concept or skill, students make four different representations, one in each fourth of a piece of paper or section of a personal whiteboard, and so on, that can be a contextual example, verbal example, graph example, and a symbolic example. The four labels may change to reflect the restrictions of the prompt.
Student work samples	Analyze student work samples for tasks you have chosen to identify understanding or misconceptions.
Exit tasks	An "exit task is a capstone problem or task that captures the major focus of the lesson for that day or perhaps the past several days and provides a sampling of student performance" (Fennell et al., 2017, p. 109).

Besides informing instructional decisions, another important part of formative assessment is that you can give feedback on what you have observed. Quick responses and immediate feedback ensure that students progress with a stronger sense of identity because they are reassured in their thinking and are not carrying forward misconceptions.

ASSESSMENT

How Can I Use Information From Formative and Summative Assessments?

129

Summative assessment	Formative assessment
Given whole class at specified times (tests, quizzes, projects) or as a large-scale assessment (state test, ACT, SAT)	Ongoing throughout class
	Diagnostic
Evaluative	Used to make immediate instructional decisions in real time as well as outside of class
Used for district and state evaluation and accountability	

The best explanation I ever heard of summative vs. formative assessment is summative assessment is assessment *of* learning and formative assessment is assessment *for* learning.

—SEVENTH-GRADE TEACHER

Notes

How Do I Support My Students With Feedback?

There are multiple ways of helping students evaluate how they are performing in your math class. Traditionally, grading has been one of those ways. But grading is not just about assigning a letter grade or number to student work. Often, an intended purpose of grading student work is to provide feedback so that students can learn from their mistakes and assess their progress toward attaining the mathematical learning goals. However, whether it be on a brief formative assessment or a large-scale summative assessment, research shows that once a grade is assigned, students pay little attention to comments on their work, thereby limiting their opportunity to keep thinking and keep learning. Thankfully, there are several alternative options to consider for providing feedback besides grades. Consider your own experiences with being graded or receiving feedback on your own work and what you learned from it (or didn't learn).

GOOD VERSUS POOR FEEDBACK

Not all feedback is good. Is your feedback too vague? Are you precise in the praise you give and in the questions you ask to move students forward? Here are some examples of what makes good or poor feedback.

Good feedback is . . .	Poor feedback is . . .
Immediate	Delayed or not given at all
Provided regularly and consistently	Sporadically given
Focused on questions that help the student think about next steps	Focused on telling students what to do next to solve a problem
Based on the work you see	Based on what you wish to see and funnels student toward a specific solution
Specific to next steps	Vague
Personalized to the student	Generic to the whole class, such as, "Many people tried this."
Offered by both teachers and peers	Only offered by the teacher

HOW CAN I PROVIDE FEEDBACK?

Here are some tips on ways to make sure you're offering feedback that meets the "good" criteria listed above.

Tip 1

While students are working, walk around the room and give verbal feedback frequently. Keep track of how students are doing using a monitoring tool. Give feedback on both correct and unfinished solutions.

A monitoring tool should be simple, so you can easily keep track of what your students are doing. This table, for use while students are collaborating, has

- a column for the strategies you anticipated students would use for the task,
- a column for who is using this strategy (group numbers you have assigned make this easier to use),
- a column to note how the students were applying the strategy, and
- steps you took as a teacher for the groups (feedback, question to get students on a new path).

This flows quickly from your lesson plan, as you have worked out the task and thought about anticipated solution paths (see Anticipating, p. 102).

Monitoring Tool

Strategy	Who	Notes on discourse/ progress	Teacher steps
Other strategy 1			
Other strategy 2			

Tip 2

Write comments on solutions, but do not assign a grade. When doing this, it is not necessary to give feedback on every solution.

Tip 3

Consider "highlighter feedback" (Paape, 2016). For this method pick four colors of highlighter. Then assign one color for each type of comment: for example, pink means minor computational error but reasoning is correct; orange means major computational or reasoning error that caused the solution to be incorrect; blue means there is a misunderstanding about a reasonable solution process, and green indicates a good solution. Since your goal is student understanding, allow students to revise and submit their final solution(s).

> Sometimes I pregrade tests for students. They turn in their test on day 1, and I go through and circle incorrect answers. They could be small computation mistakes or significant misconceptions, but the only thing students know is that something is wrong with the circled answer. On day 2, I pass the tests back and students have the opportunity to fix those problems for full credit.
>
> —SEVENTH-GRADE TEACHER

Tip 4

Use a rubric when grading student reasoning. Rubrics ensure that you have consistency while you are looking at student work as well as feedback built into the categories. You may share the rubric in advance so students have an idea of what your goals are, or you may just share the categories. A rubric may be used to show a student how they meet the goals, to identify good work, and to offer specific comments targeted at student reasoning and aimed at the student correcting their error(s). A rubric may be used as a final assessment for a task, for which the solutions should be shared so the students may reflect on their work.

ASSESSMENT

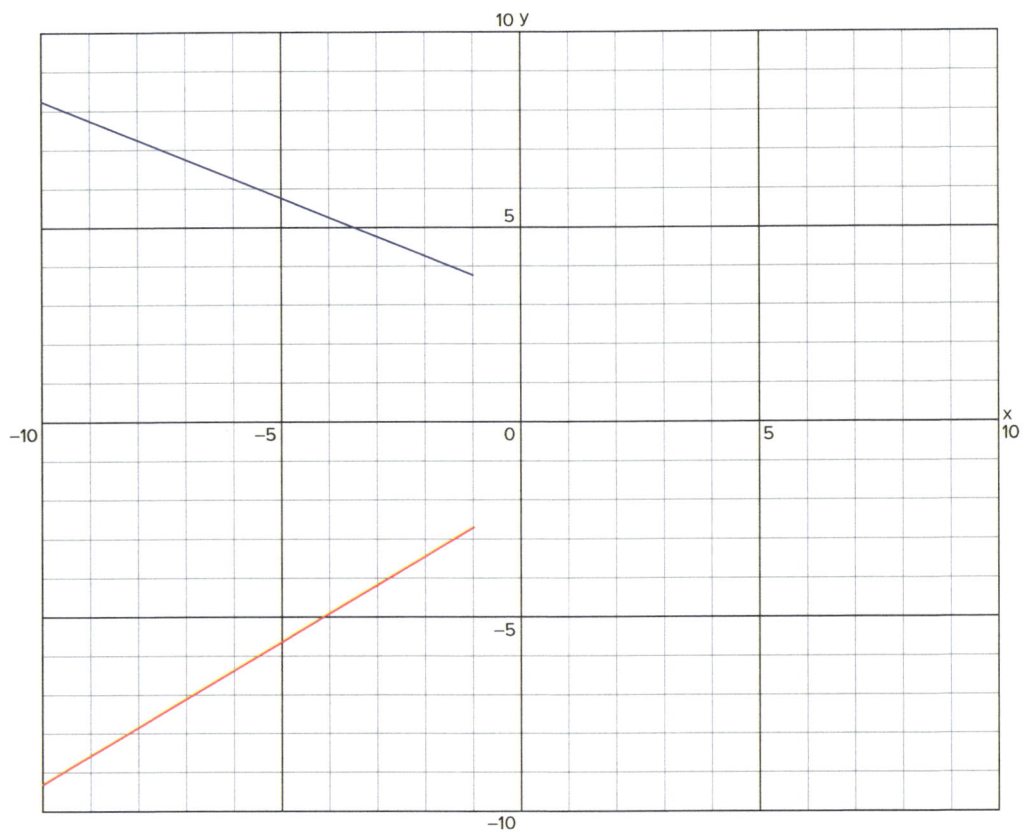

Graph task: Sample rubric

This task requires: **Interpreting a graph for rate of change (with explanation)** **Determining the point of intersection (with explanation)** **Interpreting the solution in the context of the problem**	Points	Section points
Finds the correct rate of change for line	1 point	4 points
Blue line:	1 point	
Red line:	2 points	
Strategies: Draws slope triangle on the graph, uses ordered pairs to determine the slope, makes a table, and finds the rate of change. May share only when work is finished.		
Finds the coordinates of the *y*-intercepts	1 point	4 points
Blue line:	1 point	
Red line:	2 points	
Strategies: Makes a table and finds the values where *x* is zero, extends the graphs and finds the *y*-intercepts, writes equations, and solves algebraically. May share only when work is finished.		
States: The lines	1 point	2 points
Strategies: Makes a table and finds the values where *x* is zero, extends the graphs and finds the intersection, writes equations, and solves algebraically. May share only when work is finished.	1 point	
Total		10 points

Tip 5

Peer feedback is useful for skill practice or tasks that do not have multiple solutions or pathways. Students compare their work to see what they have missed and can then discuss with their grading partner or as part of a larger/whole-class discussion.

Tip 6

For skill and algorithmically based checking for understanding, homework, or in-class work, consider asking students to explain their reasoning for an incorrect answer. You may uncover conceptual understanding more efficiently with this method. A variation on this is allowing students to revise work based on your feedback.

Tip 7

Avoid making assumptions about what the student has done. Your initial response to a solution may be to take over the student's thinking and to share about what you *think* the work shows. For example, a student may be solving a task involving a numeric solution. On seeing the correct answer, with no work, it is tempting to say, "Good job!" and to move on. However, you must ask the student about their solution process to ensure that their answer was produced by correct reasoning.

Tip 8

Provide feedback that the student can take action on. "What do you think would happen if you tried adding the angles in the triangle?" or "What would a manipulative model of your equation look like?" are examples of asking a question that presses the student to keep working but that doesn't say, "this is correct" or "this is wrong, so do this." Your question can make it clear that the solution needs to be reconsidered, but there is no specific set of steps to correct the problem.

Great Resource

For questions about grading and equity, see Joe Feldman's *Grading for Equity* (2019).

ASSESSMENT

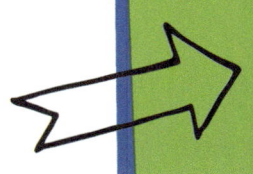

How Do I Analyze Classroom Data?

Classroom data are classroom information or artifacts that teachers can access and use to learn about their own or others' teaching practice. Usually, the term *data* is reserved to describe student-produced data such as classwork, homework, or assessment responses (see Assessment, p. 128). This is an important component of classroom data that can shed light on what students understand at a specific moment in time but is limited in telling the whole story of the classroom. Seeking other sources of classroom data is important for better understanding of how what students encounter in the classroom affects what they learn.

WHAT COUNTS AS CLASSROOM DATA?

There are three categories of data in a classroom. The figure below shares these categories and two examples of each.

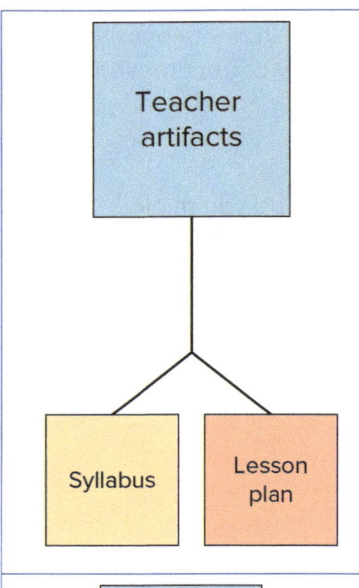

Teacher-produced data often highlight what teachers believe about their content area (Stipek et al., 2001) and their students' future goals or careers (Sztajn, 2003). Teacher artifacts are teacher-produced documents that communicate what and how learning takes place in the classroom. These artifacts include things such as rules and expectations, syllabi, lesson and unit plans, pacing guides, posters, and books on display in the classroom.

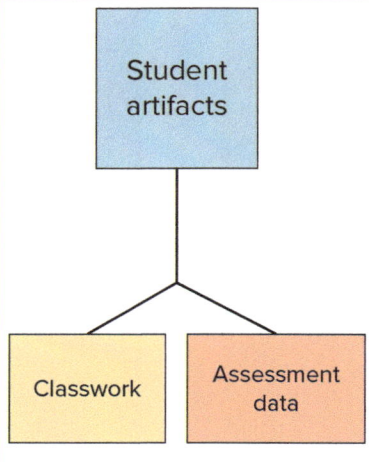

Student artifacts are student-produced documents that communicate what a student knows and understands at a single point in time. These artifacts include things such as classwork, homework, and assessment responses. Student-produced data show the teacher and students what learning has taken place (see Assessment p. 128). Student-produced data might be seen after a task has been solved or during the reasoning process on whiteboards or worksheets, or they might be shared verbally.

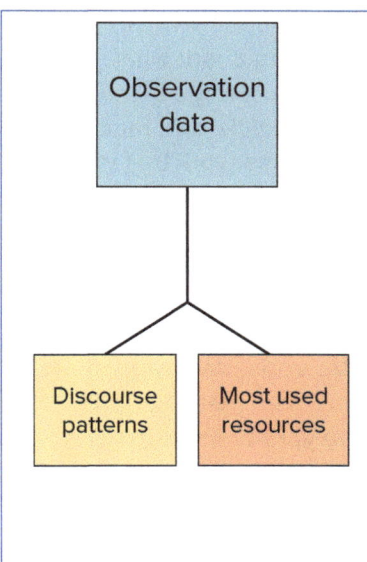

Observation data refer to anything that can be seen and includes who raises their hand first, how papers are returned to students, how students are commonly seated, and what information is written on the board. These data can clue visitors in to what the culture and norms of the classroom are. Observation data can support the teacher in learning how classroom norms and activities are taken up in the learning community. For example, if one of the classroom norms is that students ask questions of their elbow partners before talking with the teacher, an observer might record how many times students follow that norm compared with the number of times students ask the teacher first when they have questions. These data are easiest to collect with another set of eyes, whether they be a video camera or a colleague. It is difficult to manage the work of learning in the classroom and collect observation data simultaneously.

HOW DO I DECIDE WHAT DATA TO COLLECT?

Teachers should collect their data based on what they aim to learn about in their classroom. Each form of data that exists n the classroom tells one part of the story of the classroom. For example, a teacher might be interested in learning about who volunteers to share their thinking most often in class and what contributes to their comfort. To learn about this, a teacher could

- A create a seating chart for the day and record tally marks based on the number of times a student raises their hand to share their thinking with the class,
- B focus on one group during the lesson and record tally marks for the number of times different students in the group ask questions or share their thinking with other group members,
- C give students a survey to ask them about how comfortable they are sharing their thinking in large or small group settings, or
- D focus on the volunteering tendencies of their highest or lowest achievers.

Each of these data points provides information about how comfortable students feel about sharing their thinking and the circumstances that make sharing easier.

HOW CAN I ANALYZE THE DATA I HAVE COLLECTED?

Once you have decided on what you want to learn, you need to develop a plan for data analysis. You can follow these tips to support you in using the data you have collected.

Tip 1 | *Make sure you have just enough data*

One of the biggest misconceptions about data analysis in a classroom is that you have to analyze all the data you have access to in order to learn about your area of interest. This is not true and can result in you feeling overwhelmed. Instead,

ASSESSMENT

narrow your scope so that you are looking at enough data to support you in learning about your practice. For example, if you are wondering about what your students understand of a recently taught topic, you can select a range of student work to analyze. This might be reading through half of the exit tickets at random or purposefully selecting a set of students' work, including some work that was successfully completed or partially completed or where students demonstrated significant struggle (Blythe et al., 2015).

Tip 2 Find some teachers to analyze your data with you

You will learn more and more deeply if you share your classroom data with others. Teachers outside of your classroom may see things in your data that you don't notice. This larger input from others can help you improve your practice in ways that you may not have initially considered (Blythe et al., 2015).

Tip 3 Use a protocol to make sure the conversation stays on track

Structured conversations about classroom data may seem constricting, but it can be scary to share your classroom data with others. Protocols are tools that teachers can use to narrow data discussions and keep the conversations focused on what you want them to learn.

Notes

What Is the Role of Notes?

Note taking can be a useful practice to anchor the reasoning and thinking work that takes place in classes based on problem solving. Students learn math by doing math, and note taking is one way for students to summarize and consolidate their learning for the day. Notes also maintain a record of what was learned for future reference. When planning a note-taking portion of your lesson, consider these tips.

Tip 1 | *Notes should be taken by students and for students*

- Notes should be meaningful to the students. They should be a consolidation of what the students have learned and should be in their own words.
- Notes should give students the opportunity to process and organize information learned.

> I used to define good note taking as when students copied what I wrote exactly. Now I realize it made me feel good but it didn't help the students learn the material.
>
> —HIGH SCHOOL TEACHER

Tip 2 | *Notes should be taken to summarize learning*

- Notes are best taken *after* the collective summarization of learning. There are some pieces of information, such as standard notation, that just need to be told to students by the teacher and should be included in notes.
- Give time and structure during class to allow students the opportunity to write down the main points for later reflection/review.
- After taking individual notes, have students work in a group to make sure all the main points were captured.

Tip 3 | *Notes are a resource to support students in accessing their prior knowledge*

- Notes should include solution paths and written explanations of student reasoning, so that reviewing the notes supports students in remembering and making sense of the math.
- Lesson plans should have time set aside for students to revisit notes that they have taken in the past. The typical curve of forgetting newly learned information shows that most students forget material if it is not reviewed frequently.
- Research shows that if we want our students to remember more of what they learn in class, it is better to have them take notes than it is to not have them take notes (Kiewra, 2002)

Tip 4 | There are all sorts of ways to organize notes

- Many schools use an AVID focused note-taking strategy, which recommends five phases of note taking.

Step 1	Taking notes	Create the notes—record notes in a structured format
Step 2	Processing the notes	Think about the notes—underline, highlight, circle, or star main ideas
Step 3	Connecting thinking	Think beyond the notes—connect to previous learning and/or identify points of confusion
Step 4	Summarizing and reflecting on learning	Think about the notes as a whole—create a summary of the big picture
Step 5	Applying learning	Use the notes—revisit your notes at a later date

- Using graphic organizers is useful to provide a preview of what notes should look like. Cells of given size help with brevity and could also include pictures, examples, nonexamples, characteristics, vocabulary, procedures, and definitions. This also brings clarity as to important aspects you as the teacher want included in their notes.
- Fill-in-the-blank notes or guided notes provide a copy of notes to students, with certain keywords or phrases omitted. This allows teacher control over what gets written down without the need for students to copy notes directly off the board.
- Word banks provide visible vocabulary, key terms, names, and concepts that students can use to develop complete notes.
- Composition or spiral notebooks are typically places for students to record notes and practice sample problems. Foldables and paper cutouts are often included in the notebooks as well as in colorful visuals. Digital notebooks have become increasingly more popular for technology-infused classrooms.

Tip 5 | Students should be taught how to take notes

- Encourage students to include pictures and worked examples with annotation. Worked examples are useful to show the *how*, and annotation is useful to show the *why*.
- Students should be given the specific question(s) used to create worked examples. Questions that capture all the subtleties you want included yet don't get complicated with nuances work best.
- Show samples of notes, including exemplars, and have conversations about what would be most beneficial to include in notes and why.

Answers to Your Biggest Questions About Teaching Secondary Math

- After a month, give an assessment that requires their notes. This is a great way for students to see what would have been useful for them to include as well as showing them how easy it is to forget material that they thought they would remember.

Tip 6 | How you organize the class discussion and your board work matters

- By modeling effective organization of material discussed and learned, students can use the board to more easily take notes to summarize their learning.
- Student solutions and strategies should be recorded in such a way that they keep a record of the mathematical details from the lesson that both organizes student thinking and captures mathematical connections that were discussed. (Japanese teachers refer to the use and organization of the chalkboard as *bansho* or board writing.)

Notes

ASSESSMENT

WHERE DO I GO FROM HERE?

As we look at the last chapter in this book, we realize that good teaching is complex. It can feel overwhelming to be responsible for the mathematical success as well as the safety and well-being of so many students. This book sought to answer some of the most important questions teachers ask. Yet there is always more to learn. What we have shared with you has evolved over time, and your practice will likewise evolve as you gain experience and learn more as a professional. Great mathematics teaching is a journey, not a destination. Your own learning is never finished as you work to increase the impact you have on your students. We hope you embrace this opportunity to be a lifelong learner and enjoy this journey you are on. Teaching is a worthy and rewarding profession that is always full of variety and lets you share your love of math and your love of learning. It allows you the honor and privilege of preparing your students for their futures and the opportunity to inspire and make a difference in their lives.

An important step for your growth is reflecting on your practice. Part of this can be a self-assessment. Ask yourself the following:

- ☐ **What is going well in my classes?**
- ☐ **Is my mathematical community growing toward what I envision?**
- ☐ **What one new idea/process/routine do I want to try? Why do I think that would benefit my students?**
- ☐ **What do I want to change? What plans do I have to affect that change?**
- ☐ **Am I satisfied with my communications with**
  - ☐ **my students?**
  - ☐ **my colleagues?**
  - ☐ **the caregivers of my students?**
  - ☐ **administrators?**
- ☐ **And, if not, what can I try to improve my communications?**

Along with self-reflection, you may be wondering about other next steps. We hope we have answered all your concerns about those with the final question and with our list of resources and references. Your career is a lifelong journey of learning and growing. Good luck as you continue on your path.

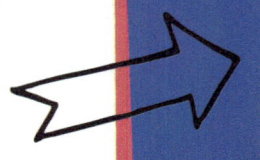

What Activities Can I Pursue to Learn and Grow?

As a teacher, you will be learning new content, skills, and pedagogy as you progress throughout your career. That is ok. No teacher ever knows it all. Every time you have the opportunity to grow in your teaching, don't pass it by. The power of strengthening both your content knowledge and your teaching knowledge will make you a better teacher and model to your students that learning is a lifelong skill. Here are some options you might pursue depending on what works best for you.

SEEK OUT PROFESSIONAL DEVELOPMENT

Look at options presented for you, and then attend the events that feel like a good fit.

Great Resources

AP Workshops, curriculum workshops

- *Within your school*
  - Attend professional development sessions led by in-school personnel.
  - Create a personal learning community or book study with a group of colleagues.
  - Attend regular staff and department meetings.
- *Within your district*
  - Ask about in-district professional development sessions across schools.
  - Attend in-district offerings according to your interests and needs.
- *Within your state or region*
  - Look for mathematics conferences in your area, including in-person and virtual meetings. Be mindful because such options may require registration and travel services.
  - Attend workshops according to a specific content.
- *Online or further away*
  - Join summer residential programs.
  - Search for options online.
  - Join local, state, and national mathematics organizations. Read their journals and attend their conferences.

TAKE COLLEGE COURSES

If you are a teacher who is new to mathematics or have transitioned to a grade or content area that is less familiar to you, taking additional courses again can refresh your memory on learning specific topics or extending your knowledge. The flexibility and growing options available online and in-person make it easier to take courses. Your work may focus on mathematics or pedagogy and new approaches for your students. One advantage of your learning is that you can experience with new eyes what it feels like for students to learn. Class fees and books are sometimes covered by scholarships, fellowships, or your district. Investigate if you can have your coursework partially or fully funded.

COMPLETE YOUR OWN HOMEWORK

If you are teaching a new course, keep refreshing your memory by completing the homework before you assign it. Practicing with homework will make you feel more confident in the topics students are learning, help you develop the learning arc for your goals, and help you see connections to other parts of the curriculum. It will also let you anticipate where students may be struggling or where they may need more support.

> As a general rule, however much time I spend doing an assignment it takes the students about four times that amount.
>
> —ALGEBRA TEACHER.

After you complete your homework assignments, you can choose to share your solutions as guides for students to review at home or in your class as part of a routine (see Homework p. 63).

ATTEND PROFESSIONAL CONFERENCES

Look for local, state, and national conferences to attend. Ask your colleagues about any conferences that they recommend and how they got funds to attend. Some conferences are free, and others provide limited funding to participants who apply in advance or present a workshop. Many conferences have a theme or specialization for a particular purpose, so it pays to research which ones best meet your current needs. When you return from the conference, try to find ways to implement what you have learned whether small or big (see Lesson Planning, p. 51). Ask in advance for time to share your learning with your colleagues at a team meeting. Conferences also provide a unique experience of gathering teachers from wide backgrounds. When you attend, take the time, even during lunch, to be brave and talk to new people so you can network.

> When I go to conferences many times I learn more from the other teachers attending than I learn at the presentation itself.
>
> —CALCULUS TEACHER

READ BOOKS AND ARTICLES

Keep yourself up to date by reading books or articles on content or pedagogy. Read in the evenings or during the summers. You can even organize a book club with your friends or colleagues to hold yourselves accountable. If you take public transit or drive to school, you can listen to audiobooks instead to optimize your time on your commute. Your school may have SSR (Silent Sustained Reading) time during a specific time of the day or week, and you can model for your students how to participate actively by reading your book then.

Great Resources

How to find books? [1] Ask for recommendations from colleagues. [2] Sign up for journals (virtual or in physical print) as part of a professional membership. [3] Look online to find the latest books from your favorite authors. [4] Visit a local or virtual bookseller or library (including electronic options such as Libby and Overdrive). [5] See our list of books at the end of this section.

TALK TO OTHER TEACHERS

Organize time for yourself to collaborate and talk with other teachers as often as possible, but at least weekly or monthly. You can have lunch in the teacher's lounge or organize a weekly lunch with a math teacher near you. It does not always have to be a structured meeting with an agenda. Having informal lunch dates with other teachers gives you the opportunity to ask a question that you are wondering about. For example, your students may have asked you, "Why does a negative number multiplied by a negative make a positive?" You gave your student a response but were not satisfied with how you justified the idea. Talking to other teachers about their approaches may provide some insight into how to go back to that same student or address the concept in the future. Formal meetings with your department or personal learning communities are equally important. These meetings allow your team to discuss topics you need to address as a group.

PARTICIPATE IN A FELLOWSHIP

Look for and apply to teaching fellowships. Fellowships combine professional development and ongoing support throughout the school year with teachers and leaders outside of your school. Each one has different requirements or expectations, so be sure there is no conflict with your personal life.

CREATE COLLABORATIVE RELATIONSHIPS

Teachers who collaborate with colleagues inside and outside their school are more effective (NCTM, 2014). When teachers collaborate, students benefit from common assessments, projects, lessons, and more. Some schools might offer structural time frames for teachers to collaborate during the school day. For example, two teachers, both teaching geometry, might have a common prep period to plan together. Other schools might have teams of teachers to meet monthly and discuss common grading practices or assessments. Below are some examples of collaboration.

- Interact with others to expand your vision of what good teaching and learning looks like.
- Work together to create tasks, projects, assessments, lessons, and so on.
- Problem solve together to offer different perspectives.
- Create supportive relationships with colleagues who listen and care.

Social media can be a great platform to collaborate! Follow teachers on Twitter like #mtbos or #iteachmath.

—FLORIDA MIDDLE SCHOOL TEACHER

Notes

How Can I Reflect on My Teaching?

One of the hallmarks of growing as a teacher is the ability to reflect on your own practice and to reach out to your peers so you can learn from their feedback and from observing them.

Here are some tips for reflection:

- Discuss with your peers
  - what you notice and wonder about in each other's classes and
  - what you saw that went great and what you want to work on for change, which are important parts of working with peers.
- Be kind to yourself! We all tend to be our own greatest critic. No lesson is perfect. Look for the good first, then focus on one or two areas for growth.
- Do not try to change everything in your practice at once. That would be a herculean task and one that will cause you and your students undue stress.
- Your district has administrators (and maybe academic coaches) who are there to help you as well. With all your peers, administrators, and coaches,
  - listen to their comments with an open mind,
  - do not take criticisms personally (the goal for everyone is to improve education for all our students), and
  - ask questions about next steps, clarifications, and what they recommend.
- Be aware that you must be true to yourself. Different personalities interact with students in different ways: what works for someone else may not work for your personality.

LEARN FROM MATH COACHES

All professionals, even experienced teachers, can benefit from content-focused instructional coaching (NCTM, 2014). Talk to your administrator or department head to find what math coaching options are available at your school or district. A math coach dedicates time to visit your classroom and works with you to set personal goals for your teaching practice. They structure and organize time for teachers to actively find ways to improve.

> My math coach gave me the time and space I needed to investigate and improve my teaching. I was able to analyze the data they collected in my classroom and incorporate strategies so more of my students participated in classroom discussions.
>
> —ALGEBRA TEACHER

PEER OBSERVATIONS

Find out if some of your colleagues are willing to let you visit their classrooms. Then, ask your principal about what will be necessary for that to happen (Will you need your own class to be covered to be able to do it? Are there any restrictions in your building about peer observations?). It may be helpful to visit teachers in other grades (or even other subjects!), especially if you can visit classes that precede and that follow your class in the pathway at your school. This can help you see new approaches to structure, management, routines, and how students interact. Pay equal attention to how the students act (Are they discussing math among themselves? Are they taking the initiative to persevere when problem solving?) as well as to how the teacher acts (How do they ask questions? Are they a facilitator?).

Before you start the visit, try to talk to your colleague about their goals for learning for the day and how they anticipate knowing if the goals were achieved. Here are some things you can look for during a class visit:

- How is the room set up?
  - Where are math tools placed for access to students?
  - What displays are there?
  - How is the classroom arranged?
- How do students interact with each other?
  - How do they work together in groups?
  - What is whole-class instruction like?
- How does the teacher monitor student thinking and work?
- What are the routines in the class?
  - How does the class start?
  - How does class end?
  - How are tasks distributed?
  - How are students placed into groups?
- What questions are asked by the teacher and how are discussions facilitated?
- Who is doing the math? Are all students represented and able to communicate?
- What does the teacher do to encourage productive struggle?

RECORDING WHAT YOU SEE IN A PEER VISIT

It is helpful to have a monitoring tool for your visit, just as it is helpful for working with your students. You need something to record on. (Do not assume you will remember what you see!) You may have a list of key "look-for" items listed, so you can write under each of those. Another option is to look for what students are doing and to connect that to corresponding teacher actions. When you reflect afterward, create a list of things you want to try or that you are reaffirmed you should continue to do. Here are two possible monitoring tools that have suggestions of how to organize what you may be looking for.

What I *noticed*	What made the noticing *stand out*	What I *wonder* about

You may want to focus specifically on student actions.

What I *noticed* students doing	What *teacher actions* influenced that	I want to try this in my class

Schedule time to talk with the teacher as soon as possible.

It is useful to have peers visit your classroom with the same monitoring protocol and follow-up discussion. You can both learn and grow by sharing your insights, observations, and wonderings about each other's classes.

RECORD YOURSELF AND VIEW THE TAPE WITH YOUR MONITORING TOOL

It is a bit frightening to record your own class. However, if you watch the tape with your monitoring tool and try to avoid just looking for "the bad things," this will help you see exactly what you do and then help you reflect about what you want to keep and what you might change. When you are teaching, there is so much going on, it is difficult to keep track of everything, let alone remember things when you have time to pause and reflect. Recording a lesson or part of a lesson gives you a way to see what you did and maybe to see some things you hadn't noticed. Pay specific attention to how you ask questions, how students are taking part in discussions, and so on.

> One of the most powerful things I did was to videotape myself teaching. As I watched it, I realized I was talking too much! I needed to make a better effort to let students be heard, not me.
>
> —HIGH SCHOOL TEACHER

LEARN FROM STUDENTS

Students are a main source of information and feedback when it comes to reflecting on your practice. Dedicate time and space in your class to ask your students to reflect and evaluate your teaching. Learning what your students have to say will help you actively reflect on your practice.

> I had an awful day, and my students looked more confused than ever. I decided to read their evaluations. I was surprised to read many students appreciated my lessons. They found my teaching style motivating. I felt their energy and found strength to remind myself how to make my lesson better for the next day.
>
> —HIGH SCHOOL TEACHER

Notes

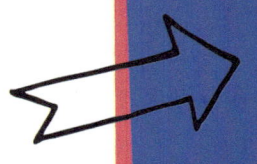

What Resources Should I Use to Learn and Grow?

There is an endless supply of resources beyond your textbook series and colleagues to grow in your professional expertise. First and foremost, you want to rely on the learning resources your school and district put forward. In addition to those, we offer the following.

ONLINE RESOURCES: WEBSITES, WEBINARS, AND BLOGS

The list of our favorite websites, blogs, and other online resources would add several pages to this book. There are specific places in the book in which relevant online resources are noted (sometimes by title, sometimes with the URL). When you search on your own, keep in mind that the internet doesn't have a process to vet the resources you may find and that you need to use a critical eye on possible resources. You can do this as follows:

- Keep a list of reliable resources that you already know and compare new resources with the items on the list.
- Read online reviews to see what professional organizations, mathematics educators, and other teachers say about a resource and to see what they recommend.
- Talk to colleagues (in your school or online) about their discoveries and experience with resources.
- If something seems too good to be true, it probably isn't as good as your first impression. Quick fixes, problems that are "fun" but that don't have clear goals or connections to what you are teaching, need to be carefully considered as to their suitability and as to how you will need to adapt them to your students' needs.

BOOKS

Great resources that connect to the questions in this book have been highlighted throughout. But those are just the start. This is a list of professional works and supplemental classroom resources that we have found invaluable over the years. These works are popular across the country. It is likely that your team, department, or district office has copies you can use. It's important to note that this list is not exhaustive. New resources are published every year, so we encourage you to continue seeking new ideas as you continue to grow your practice.

CHAPTER 1: HOW DO I BUILD A MATH COMMUNITY?

Identity, agency, and mindset	
Hammond, Z. (2014). *Culturally responsive teaching and the brain: Promoting authentic engagement and rigor among culturally and linguistically diverse students.* Corwin.	Boaler, J. (2015). *Mathematical mindsets: Unleashing students' potential through creative math, inspiring messages and innovative teaching.* Jossey-Bass.
Horn, I. (2017). *Motivated: Designing math classrooms where students want to join in.* Heinemann.	SanGiovanni, J., Katt, S., & Dykema, K. (2020). *Productive math struggle: A 6-point action plan for fostering perseverance.* Corwin.
Jansen, A. (2020). *Rough draft math: Revising to learn.* Stenhouse.	Horn, I. (2012). *Strength in numbers: Collaborative learning in secondary mathematics.* NCTM.
Dance, R., & Kaplan, T. (2018). *Thinking together: 9 beliefs for building a mathematical community.* Heinemann.	Jones, S. M. (2019). *Women who count: Honoring African American women mathematicians.* American Mathematical Society.

Equity in mathematics	
Civil, M., Crespo, S., & Fernandes, A. (2017). *Access and equity: Promoting high-quality mathematics in grades 6–8.* NCTM.	Civil, M., Crespo, S., & Fernandes, A. (2018). *Access and equity: Promoting high-quality mathematics in grades 9–12.* NCTM.
Seda, P., & Brown, K. (2021). *Choosing to see: A framework for equity in the math classroom.* David Burgess Consulting.	Snyder, S. C., & Fenner, D. S. (2021). *Culturally responsive teaching for multilingual learners: Tools for equity.* Corwin.
Feldman, J. (2019). *Grading for equity: What it is, why it matters, and how it can transform schools and classrooms.* Corwin.	Berry, R. Q., Conway, B. M., Lawler, B., & Stanley, J. (2020). *High school mathematics lessons to explore, understand, and respond to social injustice.* Corwin.
Padilla, A., Mason, E., & Sheldon, J. (2019). *Humanizing disability in mathematics education: Forging new paths.* NCTM.	

CHAPTER 2: HOW DO I STRUCTURE, ORGANIZE, AND MANAGE MY MATH CLASS?

Routines	
SanGiovanni, J., & Milou, E. (2018). *Daily routines to jump-start math class, middle school mathematics: Engage students, improve number sense, and practice reasoning.* Corwin.	Milou, E., & SanGiovanni, J. (2018). *Daily routines to jump-start math class, high school: Engage students, improve number sense, and practice reasoning.* Corwin.
McCoy, A., Barnett, J., & Combs, E. (2013). *High-yield routines for grades K–8.* NCTM.	Kaplinsky, R. (2019). *Open middle math: Problems that unlock student thinking, grades 6–12.* Stenhouse.
Kelemenik, G., Lucenta, A., & Creighton, S. J. (2016). *Routines for reasoning: Fostering the mathematical practices in all students.* Heinemann.	Humphries, C., & Parker, R. (2015). *Making number talks matter: Developing mathematical practices and deepening understanding, grades 3–10* (Illustrated edition). Stenhouse.

(Continued)

Routines	
Parker, R., & Humphries, C. (2018). *Digging deeper: Making number talks matter even more*. Steinhouse.	Luzniak, C. (2019). *Up for debate! Exploring math through argument*. Stenhouse.

Instructional practice	
Zager, T. (2017). *Becoming the math teacher you wish you'd had: Ideas and strategies from vibrant classrooms*. Stenhouse.	Liljedahl, P. (2021). *Building thinking classrooms in mathematics*. Corwin.
Kanold-McIntyre, J., Larson, M., Briars, D., & Kanold, T. D. (2012). *Common Core mathematics in a PLC at work, grades 6–8*. NCTM & Solution Tree.	Zimmerman, G., Carter, J., Toncheff, M., & Kanold, T. D. (2012). *Common Core mathematics in a PLC at work, high school*. NCTM & Solution Tree.
Nolan, E., Dixon, J. K., Roy, G. J., & Andreasen, J. B. (2016). *Making sense of mathematics for teaching: Grades 6–8 (unifying topics for an understanding of functions, statistics, and probability)*. Solution Tree.	Nolan, E., Dixon, J. K., Safi, F., & Haciomeroglu, E. S. (2016). *Making sense of mathematics for teaching high school (understanding how to use functions)*. Solution Tree.
Moore, S. D., & Rimbey, K. A. (2021). *Mastering math manipulatives: Hands-on and virtual activities for building and connecting mathematical ideas*. Corwin.	Bush, S. B., Karp, K., & Dougherty, B. J. (2020). *The Math pact, middle school: Achieving instructional coherence within and across grades*. Corwin.
Dougherty, B. J., Bush, S. B., & Karp, K. (2020). *The math pact, high school: Achieving instructional coherence within and across grades*. Corwin.	Almarode, J. T., Fisher, D., Assof, J., Hattie, J., Frey, N. (2018). *Teaching mathematics in the visible learning classroom, grades 6–8*. Corwin.
Almarode, J. T., Fisher, D., Assof, J., Hattie, J., & Frey, N. (2018). *Teaching mathematics in the visible learning classroom, high school*. Corwin.	Schuhl, S., Kanold, T. D., Kanold-McIntyre, J., Chuang, S., Larson, M. R., & Smith, M. (2020). *Mathematics unit planning in a PLC at work, grades 6–8*. Solution Tree.
Schuhl, S., Kanold, T. D., Barnes, B., Jain, D. J., Larson, M. R., & Mozingo, B. (2021). *Mathematics unit planning in a PLC at work, high school*. Solution Tree.	Liljedahl, P. (2021). *Modifying your thinking classroom for different settings*. Corwin.
Smith, M. S., Bill, V. L., & Steele, M. J. (2020). *On-your-feet guide: Modifying mathematical tasks: Eight strategies to engage students in thinking and reasoning*. Corwin.	Hattie, J., & Zierer, K. (2020). *On-your-feet guide: Visible learning: 10 mindframes for teachers*. Corwin.
Ray-Riek, M. (2013). *Powerful problem solving: Activities for sense making with the mathematical practices*. Heinemann.	NCTM. (2014). *Principles to actions: Ensuring mathematical success for all*. NCTM.
Yeh, C., Ellis, M., & Hurtado, C. K. (2017). *Reimagining the mathematics classroom: Creating and sustaining productive learning environments*. NCTM.	Seeley, C. (2014). *Smarter than we think: More messages about math, teaching and learning in the 21st century—a resource for teachers, leaders, policy makers and families*. Math Solutions.

Instructional practice	
Wills, T. (2020). *Teaching math at a distance, grades K–12: A practical guide to rich remote instruction*. Corwin.	Van de Walle, J., Bay-Williams, J., Lovin, L., & Karp, K. (2017). *Teaching student-centered mathematics: Developmentally appropriate instruction for grades 6–8*. Pearson.

Differentiation	
Driscoll, M., Nikula, J., & DePiper, J. N. (2016). *Mathematical thinking and communication: Access for English learners*. Heinemann.	Small, M., & Lin, A. (2010). *More good questions: Great ways to differentiate secondary mathematics instruction*. NCTM.
Sheffield, L., Assouline, S., & Saul, M. (2010). *The peak in the middle: Developing mathematically gifted students in the middle grades*. NCTM.	Chval, K., Smith, E. M., Trigos-Carrillo, L., & Pinnow, R. J. (2021). *Teaching math to multilingual students, grades K–8: Positioning English learners for success*. Corwin.
National Council for Gifted Children, & Sheffield, L. (2012). *Using the Common Core State Standards for mathematics with gifted and advanced learners*. Routledge.	

CHAPTER 3: HOW DO I ENGAGE MY STUDENTS IN MATH?

Mathematics content	
Miles, R. H., & Williams, L. (2016). *The Common Core mathematics companion: The standards decoded, grades 6–8: What they say, what they mean, how to teach them.* Corwin.	Dillon, F. L., Martin, W. G., Conway, B. M., & Strutchens, M. E. (2017). *The Common Core mathematics companion: The standards decoded, high school: What they say, what they mean, how to teach them.* Corwin.
Lloyd, G., Herbel-Eisenmann, B., & Star, J. (2011). *Developing essential understanding of expressions, equations, and functions for mathematics in grades 6–8.* NCTM.	Cooney, T. J., Beckmann, S., & Lloyd, G. (2010). *Developing essential understanding of functions for mathematics in grades 9–12.* NCTM.
Sinclair, N., Pimm, D., & Skelin, M. (2012). *Developing essential understanding of geometry for mathematics in grades 6–8.* NCTM.	Sinclair, N., Skelin, M., & Pimm, D. (2012). *Developing essential understanding of geometry for mathematics in grades 9–12.* NCTM.
Lobato, J., Ellis, A. B., & Charles, R. I. (2010). *Developing essential understanding of ratios, proportions, and proportional reasoning for mathematics in grades 6–8.* NCTM.	Ellis, A., Bieda, K., & Knuth, E. J. (2012). *Developing essential understanding of proof and proving for mathematics in grades 9–12.* NCTM.
Strutchens, M. E., & Quander, J. R. (2011). *Focus in high school mathematics: Fostering reasoning and sense making for all students.* NCTM.	Graham, K. (2010). *Focus in high school mathematics: Reasoning and sense making in algebra.* NCTM.
McCrone, S. M., King, J., & Orihuela, Y. (2010). *Focus in high school mathematics: Reasoning and sense making in geometry.* NCTM.	Martin, W. G. (2009). *Focus in high school mathematics: Reasoning and sense making.* NCTM.

(Continued)

(Continued)

Mathematics content	
Shaughnessy, M., Chance, B., & Kronendonk, H. (2009). *Focus in high school mathematics: Reasoning and sense making in statistics and probability.* NCTM.	Dick, T. P., & Hollebrands, K. F. (2011). *Focus in high school mathematics: Technology to support reasoning and sense making.* NCTM.
De Aroujo, Z., Dougherty, B. J., & Zenigami, F. (2018). *Putting essential understanding of expressions and equations into practice in grades 6–8.* NCTM.	Ronau, R., Meyer, D., Crites, T., & Dougherty, B. (2014). *Putting essential understanding of functions into practice in grades 9–12.* NCTM.
Dougherty, B. J., Karp, K., Slovin, H., & Crites, T. (2017). *Putting essential understanding of geometry into practice in grades 6–8.* NCTM.	Ronau, R., Meyer, D., & Crites, T. (2014). *Putting essential understanding of geometry into practice in grades 9–12.* NCTM.
Olsen, T., Olsen, M., & Slovin, H. (2015). *Putting essential understanding of ratios and proportions into practice in grades 6–8.* NCTM.	St. Laurent, R. (2015). *Putting essential understanding of statistics into practice in grades 9–12.* NCTM.
Small, M. (2013). *Uncomplicating fractions to meet Common Core Standards in math, K–7.* Teachers College Press.	Singh, S., & Brownell, C. (2020). *Math recess.* Impress.

Mathematics practices	
O'Connell, S., & SanGiovanni, J. (2013). *Putting the practices into action: Implementing the Common Core Standards for mathematical practice, K–8.* Heinemann.	Smith, M. S., Steele, M. D., & Raith, M. L. (2017). *Taking action: Implementing effective mathematics teaching practices in grades 6–8.* NCTM.
Boston, M., Dillon, F. L., Smith, M. S., & Miller, S. (2017). *Taking action: Implementing effective mathematics teaching practices in grades 9–12.* NCTM.	

Classroom activities and tasks (6–12)	
NCTM. (2021). *Activity gems for the 6–8 classroom.* NCTM.	Morrow-Leong, K., Moore, S. D., & Gojak, L. M. (2020). *Mathematize it! Going beyond key words to make sense of word problems, grades 6–8.* Corwin.
Smith, M. S. (2020). *The on-your-feet-guide to modifying tasks.* Corwin.	

Fluency	
Bay-Williams, J., & SanGiovanni, J. (2021). *Figuring out fluency in mathematics teaching and learning, grades K–8: Moving beyond basic facts and memorization.* Corwin.	Cardone, T. (2015). *Nix the tricks: A guide to avoiding shortcuts that cut out math concept development.* CreateSpace Independent Publishing Platform.
Parrish, S., & Dominick, A. (2016). *Number talks: Fractions, decimals, and percentages.* Solution Tree.	

CHAPTER 4: HOW DO I HELP MY STUDENTS TALK ABOUT MATH AND SHARE THEIR MATHEMATICAL THINKING?

Discourse	
Smith, M. S., & Stein, M. K. (2018). *5 Practices for orchestrating productive mathematics discussions*. Corwin and NCTM.	Smith, M. S., & Sherin, M. G. (2019). *The five practices in practice: Successfully orchestrating mathematics discussions in your middle school classroom*. Corwin.
Smith, M. S., Steele, M. D., & Sherin, M. G. (2020). *The five practices in practice: Successfully orchestrating mathematics discussions in your high school classroom*. Corwin.	Kazemi, K., & Hintz, A. (2014). *Intentional talk: How to structure and lead productive mathematical discussions*. Stenhouse.
Smith, M. S., & Sherin, M. G. (2019). *The on-your-feet guide to the five practices*. Corwin.	Clarke, S. J. H. (2020). *The on-your-feet guide to partner talk*. Corwin.
Cohen, E., & Lotan, R. (2014). *Designing groupwork: Strategies for the heterogeneous classroom*. Teachers College Press.	

CHAPTER 5: HOW DO I KNOW WHAT MY STUDENTS KNOW, AND HOW CAN I USE THAT INFORMATION TO PLAN AND MOVE FORWARD?

Formative assessment	
Fennell, F., Kobett, B. M., & Wray, J. A. (2017). *The formative 5: Everyday assessment techniques for every math classroom*. Corwin.	Wiliam, D. (2011). *Embedded formative assessment-practical strategies and tools for K–12 teachers*. Solution Tree.
Wiliam, D. (2017). *Embedded formative assessment (strategies for classroom formative assessment that drives student engagement and learning)* (2nd ed.). Solution Tree.	Keeley, P. D., & Tobey, R. T. (2011). *Mathematics formative assessment: Vol. 1. 75 practical strategies for linking assessment, instruction, and learning*. Corwin and NCTM.
Keeley, P. D., & Tobey, R. T. (2017). *Mathematics formative assessment: Vol. 2. 50 more practical strategies for linking assessment, instruction, and learning*. Corwin and NCTM.	SanGiovanni, J. (2017). *Mine the gap for mathematical understanding, grades 6–8: Common holes and misconceptions and what to do about them*. Corwin.
Fennell, F., Kobett, B. M., & Wray, J. A. (2019). *The on-your-feet guide to the formative 5: Everyday assessment techniques for every math classroom*. Corwin.	

REFERENCES

Aguirre, J., Mayfield-Ingram, K., & Martin, D. (2013). *The impact of identity in K–8 mathematics learning and teaching: Rethinking equity-based practices.* National Council of Teachers of Mathematics.

American Mathematical Society. *AMS posters.* http://www.ams.org/publicoutreach/posters/posters

arbitrarilyclose. *Mathematician project.* https://arbitrarilyclose.com/mathematician-project/

Berry, R. Q., III. (2016). Informing teachers about identities and agency: Using the stories of black middle school boys who are successful with school mathematics. In E. Silver & P. A. Kenney (Eds.), *More lessons learned from research: Helping all students understand important mathematics* (Vol. 2, pp. 25–37). National Council of Teachers of Mathematics.

Bisplinghoff, B. (2017, March 30). *Norms construction: A process of negotiation.* School Reform Initiative. https://www.schoolreforminitiative.org/download/norms-construction-a-process-of-negotiation/

Black, P., & Wiliam, D. (2010). Inside the black box: Raising standards through classroom assessment. *Phi Delta Kappan, 92*(1), 81–90. https://doi.org/10.1177/003172171009200119

Blythe, T., Allen, D., & Powell, B. S. (2015). *Looking together at student work: A companion guide to assessing student learning.* Teachers College Press.

Boston, M., Dillon, F. L., Smith, M. S., & Miller, S. (2017). *Taking action: Implementing the effective teaching practices in grades 9–12.* National Council of Teachers of Mathematics.

Boston College. *Student math survey.* https://www.bc.edu/research/intasc/PDF/opd_bat_StudentPost_fall06.pdf

Cardone, T. *Nix the tricks.* https://nixthetricks.com/index.html

Centre for Justice & Reconciliation. A program of Prison Fellowship International. http://restorativejustice.org/#sthash.dNR4hqEJ.dpbs

Centre for Teaching Excellence. *Teaching problem-solving skills.* University of Waterloo. https://uwaterloo.ca/centre-for-teaching-excellence/teaching-resources/teaching-tips/developing-assignments/cross-discipline-skills/teaching-problem-solving-skills

Cohen, E., & Lotan, R. (2014). *Designing groupwork: Strategies for the heterogeneous classroom.* Teachers College Press.

Danielson, C. (2013). *Which one doesn't belong?* https://wodb.ca/

Dixon, J. K. (2020, November 17). *Just-in-time vs. just-in-case scaffolding: How to foster productive perseverance.* https://bit.ly/3ENhxEn

Dougherty, B. J., Bush, S. B., & Karp, K. S. (2020). *The math pact: Achieving instructional coherence within and across grades, high school.* Corwin/National Council of Teachers of Mathematics.

Driscoll, M. (1999). *Fostering algebraic thinking.* Heinemann.

Feldman, J. (2019). *Grading for equity.* Corwin.

Fennel, F., Kobett, B. M., & Wray, J. A. (2017). *The formative 5: Everyday assessment techniques for every math classroom.* Corwin.

Fernandez, M. L., Hadaway, N., Wilson, J. W., & Graeber, A. O. (1994). Connecting research to teaching: Problem solving: Managing it all. *Mathematics Teacher, 87*(3), 195–199. https://doi.org/10.5951/MT.87.3.0195

Finkel, D. Math for love. www.mathforlove.com

Fuson, K. C. (2003). Toward computational fluency in multidigit multiplication and division. *Teaching Children Mathematics, 9*(6), 300–305. https://doi.org/10.5951/TCM.9.6.0300

Fuson, K. C., & Beckmann, S. (2012–2013). Standard algorithms in the Common Core State Standards. *National Council of Supervisors of Mathematics Journal of Mathematics Education Leadership, 14*(1), 14–30.

Gapminder. https://www.gapminder.org/tools.

Garfunkel, S., & Montgomery, M. (Eds.). (2019). *Guidelines for assessment and instruction in mathematical modeling education* (2nd ed.). COMAP and SIAM.

Gay, G. (2000). *Culturally responsive teaching: Theory, research, and practice.* Teachers College Press.

Goffney, I., Gutiérrez, R., & Boston, M. (Eds.). (2018). *Rehumanizing mathematics for black, indigenous, and Latinx students* (Annual Perspectives in Mathematics Education, Vol. 2018). National Council of Teachers of Mathematics.

Gonchar, M. (2021, May 11). Teach about inequality with these 28 New York Times graphs. *The New York Times.*

Hamburger, A., Helft, S., & Moynihan, F. (2021, July 14). Rewriting our list of mathematicians. *Desmos.* https://blog.desmos.com/articles/new-mathematicians-list/

Hammond, Z. L. (2015). *Culturally responsive teaching and the brain.* Corwin.

Hattie, J., Fisher, D., Frey, N., Gojak, L., & Moore, S. D. (2018). *Visible learning for mathematics, grades K–12: What works best to optimize student learning.* Corwin.

Herbel-Eisenmann, B. A., & Breyfogle, M. L. (2005). Questioning our patterns of questioning. *Mathematics Teaching in the Middle School, 10*(9), 484–489. https://doi.org/10.5951/MTMS.10.9.0484

Hollins, E. R. (1996). *Culture in school learning: Revealing the deep meaning.* Lawrence Erlbaum.

Horn, I. (2012). *Strength in numbers: Collaborative learning in secondary mathematics.* NCTM.

Huinker, D., & Bill, V. (2017). *Taking action: Implementing the effective teaching practices in grades pre-K–5.* National Council of Teachers of Mathematics.

Isaacs, A. C., & Carroll, W. M. (1999). Strategies for basic-facts instruction. *Teaching Children Mathematics, 5*(9), 508–515. https://doi.org/10.5951/TCM.5.9.0508

Johnson, S. K., & Sheffield, L. J. (2012). *Using the common core state standards for mathematics with gifted and advanced learners.* National Council of Teachers of Mathematics.

Kaplinsky, R., & Johnson, N. (2016). What's open middle? *Open Middle.* https://www.openmiddle.com/whats_open_middle/

Kiewra, K. A. (2002). How classroom teachers can help students learn and teach them how to learn. *Theory Into Practice, 41*(2), 71–80. https://doi.org/10.1207/s15430421tip4102_3

Klerlein, J., & Hervey, S. (n.d.). Mathematics as a complex problem-solving activity. *Generation Ready.* https://www.generationready.com/white-papers/mathematics-as-a-complex-problem-solving-activity/

Klerlein, J., & Sheena, H. Mathematics as a complex problem-solving activity. *Generation Ready.* https://www.generationready.com/white-papers/mathematics-as-a-complex-problem-solving-activity/

Kobett, B. M, & Karp, K. S. (2020). *Strengths-based teaching and learning in mathematics: Teaching turnarounds for grades K-6.* Corwin.

Ladson-Billings, G. (1994). *The dreamkeepers: Successful teachers of African-American children* (2nd ed.). Jossey-Bass.

Lesh, R., Post, T. R., & Behr, M. (1987). Representations and translations among representations in mathematics learning and problem solving. In C. Janvier (Ed.), *Problems of representations in the teaching and learning of mathematics* (pp. 33–40). Lawrence Erlbaum.

Liljedahl, P. (2021). *Building thinking classrooms in mathematics, grades K–12.* Corwin.

Martin, A. J., & Marsh, H. W. (2006). Academic resilience and its psychological and educational correlates: A construct validity approach. *Psychology in the Schools, 43,* 267–281. https://doi.org/10.1002/pits.20149

Martin, D. (2000). *Mathematics success and failure among African-American youth: Roles of sociohistorical context, community forces, school influence, and individual agency.* Routledge.

Martin, W. G. (2009). The NCTM high school curriculum project: Why it matters to you. *The Mathematics Teacher, 103*(3), 164–166. https://doi.org/10.5951/MT.103.3.0154

Mashup Math LLC. (2019, Feb 13). *11 famous African American mathematicians you should know about.* https://www.mashupmath.com/blog/famous-african-american-mathematicians

Mehan, H. (1979). *Learning lessons: Social organization in the classroom.* Harvard University Press.

Meyer, D. *3 act tasks.* San Francisco Unified School District Mathematics Department. http://www.sfusdmath.org/3-act-tasks.html

Milou, E., & SanGiovanni, J. (2018). *Daily routines to jump-start math class, high school: Engage students, improve number sense, and practice reasoning.* Corwin.

Morrison, N., & Berlin, J. (Eds.) (2019). *Racially expansive STEM histories: Resources.* Math for America. https://www.mathforamerica.org/sites/default/files/uploads/Racially-Expansive-STEM-Histories-Resource.pdf

Multon, K. D., Brown, S. D., & Lent, R. W. (1991). Relation of self-efficacy beliefs to academic outcomes: A meta-analytic investigation. *Journal of Counseling Psychology, 38,* 30–38. https://doi.org/10.1037/0022-0167.38.1.30

National Council of Teachers of Mathematics. (1991). *Professional standards for teaching mathematics.* National Council of Teachers of Mathematics.

National Council of Teachers of Mathematics. (2000). *Principles and standards for school mathematics.* National Council of Teachers of Mathematics.

National Council of Teachers of Mathematics. (2014). *Principles to actions.* National Council of Teachers of Mathematics.

National Council of Teachers of Mathematics. (2018). *Catalyzing change in high school mathematics: Initiating critical conversations.* National Council of Teachers of Mathematics.

National Council of Teachers of Mathematics. (2020). *Catalyzing change in middle school mathematics: Initiating critical conversations.* National Council of Teachers of Mathematics.

National Governors Association Center for Best Practices, & Council of Chief State School Officers. (2010). *Common core state standards for mathematics.*

National Research Council, & Mathematics Learning Study Committee. (2001). *Adding it up: Helping children learn mathematics* (J. Kilpatrick, J. Swafford, & B. Findell, Eds.). National Academies Press. https://doi.org/10.17226/9822

Nieto, S. (1996). *Affirming diversity: The sociopolitical context of multicultural education* (2nd ed.). Longman.

PBS Learning Media. *Browse by standards.* https://bit.ly/39u6Qbc

Osler, J. RadicalMath. https://www.radicalmath.org/

Paape, Adam. (2016). Using highlighters to promote productive struggle and deepen student understanding. *Wisconsin Teacher of Mathematics. 69,* 25–29.

Perkins, A. (2017). *Mathematicians are not just white dudes.* National Council of Teachers of Mathematics Chicago Regional Session. https://docs.google.com/presentation/d/1nO6IJhmR2Z7V7D3eC5VWf7FLR3ljFmvqN_r-YhJGAaw/edit#slide=id.g23fe83d49e_0_56

Polya, G. (1945). *How to solve it* (2nd ed.). Princeton University Press.

Ramirez, G., Gunderson, E. A., Levine, S. C., & Beilock, S. L. (2013). Math anxiety, working memory, and math achievement in early elementary school. *Journal of Cognition and Development, 14*(2), 187–202. https://doi.org/10.1080/15248372.2012.664593

Reinhart, S. (2000, April). Never say anything a kid can say! *Mathematics Teaching in the Middle School, 5*(8), 478–483. https://doi.org/10.5951/MTMS.5.8.0478

Rohrer, D. (2009). The effects of spacing and mixed practice problems. *Journal for Research in Mathematics Education, 40*(1), 4–17. https://doi.org/10.5951/jresematheduc.40.1.0004

Robinson, J. Julia Robinson Mathematics Festival. https://www.jrmf.org/

Rubel, L. H., & McCloskey, A. V. (2021). Contextualization of mathematics: Which and whose world. Educational Studies in Mathematics, 107(2), 383–404. https://doi.org/10.1007/s10649-021-10041-4

SanGiovanni, J., Katt, S., & Dykema, K. (2020). *Productive math struggle: A six-point action plan for fostering perseverance.* Corwin.

SanGiovanni, J., & Milou, E. (2018). *Daily routines to jump-start math class, middle school mathematics: Engage students, improve number sense, and practice reasoning.* Corwin.

School Reform Initiative. (n.d.). *Protocols.* https://www.schoolreforminitiative.org/protocols/

Seeley, C. (2017). Unleashing problem solvers. *Educational Leadership, 75*(2), 32–36.

Silver, E., & Mills, V. (Eds.). (2018). *A fresh look at formative assessment in mathematics teaching.* National Council of Teachers of Mathematics.

Singh, S., & Brownell, C. (2019). *Math recess: Playful learning in an age of disruption.* Impress.

Skaalvik, E. M., & Skaalvik, S. (2004). Self-concept and self-efficacy: A test of the internal/external frame of reference model and predictions of subsequent motivation and achievement. *Psychological Reports, 95,* 1187–1202. https://doi.org/10.2466/pr0.95.3f.1187-1202

Smith, M. S., Bill, V. L., & Hughes, E. K. (2008). Thinking through a lesson: Successfully implementing high level tasks. *Mathematics Teaching in the Middle School, 14*(3), 132–138. https://doi.org/10.5951/MTMS.14.3.0132

Smith, M. S., Bill, V. L., & Steele, M. D. (2021, March). *On-your-feet guide: Modifying mathematical tasks (Eight strategies to engage students in thinking and reasoning).* Corwin. https://us.corwin.com/en-us/nam/on-your-feet-guide-modifying-mathematical-tasks/book272442

Smith, M. S., Steele, M., & Raith, M. L. (2017). *Taking action: Implementing the effective teaching practices in grades 6–8.* National Council of Teachers of Mathematics.

Smith, M. S., & Stein, M. K. (2011). *5 practices for orchestrating productive mathematics discussions.* National Council of Teachers of Mathematics.

Stein, M. K. (2007). *Selecting the right curriculum* (J. Reed, Ed.). National Council for Teachers of Mathematics.

Stein, M. K., & Smith, M. S. (1998). Mathematical tasks as a framework for reflection: From research to practice. *Mathematics Teaching in the Middle School, 3*(4), 268–275. https://doi.org/10.5951/MTMS.3.4.0268

Stipek, D., Givvin, K., Salmon, J., & Macgyvers, V. (2001). Teachers' beliefs and practices related to mathematics instruction. *Teaching and Teacher Education, 17*(2), 213–226. https://doi.org/10.1016/S0742-051X(00)00052-4

Style, E. (1988). *Curriculum as window and mirror. From listening for all voices: Gender balancing the school curriculum* (M. Crocco, Ed.). Oak Knoll School of the Holy Child.

Su, F. (2017). Mathematics for human flourishing. *American Mathematical Monthly, 124*(June/July), 483–493. https://doi.org/10.4169/amer.math.monthly.124.6.483

Sztajn, P. (2003). Adapting reform ideas in different mathematics classrooms: Beliefs beyond mathematics. *Journal of Mathematics Teacher Education, 6,* 53–75. https://doi.org/10.1023/A: 1022171531285

Taggart, G. L., Adams, P. E., Eltze, E., Heinrichs, J., Hohmann, J., & Hickman, K. (2007). Fermi questions. *Mathematics Teaching in the Middle School, 13*(3), 164–167.

Tomlinson, C. A. (2017). *How to differentiate instruction in academically diverse classrooms.* ASCD.

U.S. Census Bureau. (2019, August 20). Young adults and higher education. https://www.census.gov/library/visualizations/interactive/young-adults-higher-education.html

U.S. Census Bureau. (2021a, May 11). Quarterly workforce Indicators. https://www.census.gov/library/visualizations/2021/econ/quarterly-workforce-indicators.html

U.S. Census Bureau. (2021b, March 2). Women's earnings. https://www.census.gov/library/visualizations/2021/comm/womens-earnings.html

U.S. Department of Energy (2018, Aug. 29). *Women in STEM posters,* series one. https://www.energy.gov/downloads/women-stem-posters-series-one

U.S. Census Bureau. (2021c, Oct. 8). *Census Bureau 101 for students.* https://www.census.gov/programs-surveys/sis/about/students101.html

Warschauer, H. K. (2011). *The role of productive struggle in teaching and learning middle school mathematics* [Unpublished doctoral dissertation]. University of Texas.

Warschauer, H. K. (2015). Strategies to support productive struggle. *Mathematics Teaching in the Middle School, 20*(7), 390–393. https://doi.org/10.5951/mathteacmiddscho.20.7.0390

Watanabe, M., & Evans, L. (2015, November). Assessments that promote collaborative learning. *The Mathematics Teacher, 109*(4), 298–304.

Wathall, J. T. H. (2016). *Concept-based mathematics: Teaching for deep understanding in secondary classrooms.* Corwin.

Wentworth, M. (2021, March 26). *Forming ground rules (creating norms).* School Reform Initiative. https://www.schoolreforminitiative.org/download/forming-ground-rules-creating-norms/

Wiliam, D. (2015). Designing great hinge questions. *Questioning for Learning, 73*(1),40–44.

Williams, L. A., Kobett, B. M., & Miles, R. H. (2019). *The mathematics lesson planning handbook, grades 6–8.* Corwin.

Wills, T. (2020). *Teaching math at a distance, grades K–12: A practical guide to rich remote instruction.* Corwin.

Wood, T. (1998). Alternative patterns of communication in mathematics classes: Funneling or focusing? In H. Steinbring, M. G. B. Bussi, & A. Sierpinska (Eds.), *Language and Communication in the Mathematics Classroom* (pp. 67–78). National Council of Teachers of Mathematics.

Women You Should Know. *Downloadable STEM role models posters.* https://womenyoushouldknow.net/downloadable-stem-role-models-posters

INDEX

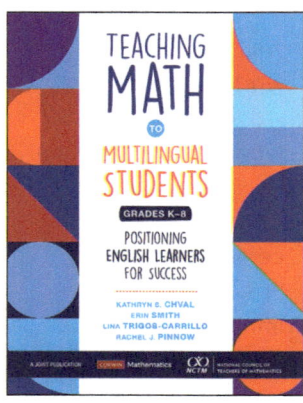

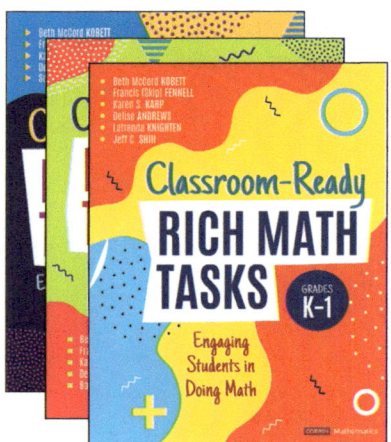

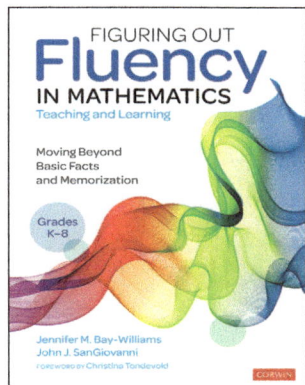
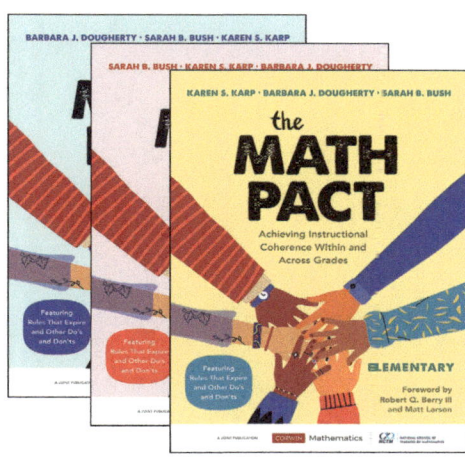
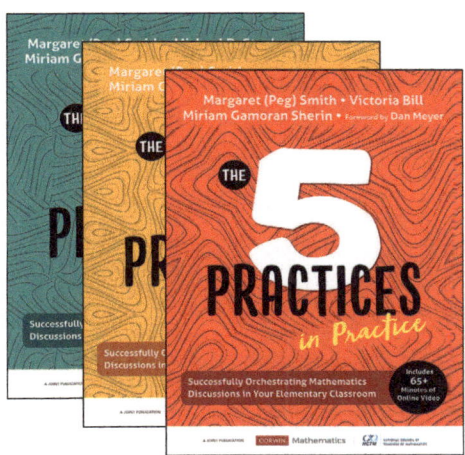

CORWIN

A SAGE Publishing Company

Helping educators make the greatest impact

CORWIN HAS ONE MISSION: to enhance education through intentional professional learning.

We build long-term relationships with our authors, educators, clients, and associations who partner with us to develop and continuously improve the best evidence-based practices that establish and support lifelong learning.